T0135562

Markus Müller-Trapet

Measurement of Surface Reflection Properties

Concepts and Uncertainties

Logos Verlag Berlin GmbH

Aachener Beiträge zur Technischen Akustik

Editor:
Prof. Dr. rer. nat. Michael Vorländer
Institute of Technical Acoustics
RWTH Aachen University
52056 Aachen
www.akustik.rwth-aachen.de

Bibliographic information published by the Deutsche Nationalbibliothek

The Deutsche Nationalbibliothek lists this publication in the Deutsche Nationalbibliografie; detailed bibliographic data are available in the Internet at http://dnb.d-nb.de .

D 82 (Diss. RWTH Aachen University, 2015)

ISBN 978-3-8325-4120-0
ISSN 1866-3052
Vol. 23

Logos Verlag Berlin GmbH
Comeniushof, Gubener Str. 47,
D-10243 Berlin

Tel.:   +49 (0)30 / 42 85 10 90
Fax: +49 (0)30 / 42 85 10 92
http://www.logos-verlag.de

Measurement of Surface Reflection Properties

Concepts and Uncertainties

Von der Fakultät für Elektrotechnik und Informationstechnik der
Rheinisch-Westfälischen Technischen Hochschule Aachen
zur Erlangung des akademischen Grades eines
DOKTORS DER INGENIEURWISSENSCHAFTEN
genehmigte Dissertation

vorgelegt von

Dipl.-Ing.
Markus Müller-Trapet
aus Düsseldorf

Berichter:

Universitätsprofessor Dr. rer. nat. Michael Vorländer
Universitätsprofessor ir. Jean-Jacques Embrechts

Tag der mündlichen Prüfung: 08. Juli 2015

Diese Dissertation ist auf den Internetseiten der Hochschulbibliothek online verfügbar.

"I'd take the awe of understanding over the awe of ignorance any day."

Douglas Adams

Abstract

Although the quality of room acoustic simulations has increased significantly in recent years, an entirely realistic result is seldom achieved in complex scenarios. Among the factors influencing the degree of realism of such simulations, the boundary conditions concerning sound reflection are considered most important as they determine the sound field to a great extent. Standardized measurement methods exist but they contain inherent uncertainties or do not always yield enough information for a correct modeling of the sound field. To ameliorate the situation, acoustic measurement techniques related to the absorbing as well as the scattering properties of architectural surfaces are investigated in this thesis.

The research is divided into two parts: the first part consists of determining the most relevant causes of uncertainty for the standardized measurement methods of random-incidence absorption and scattering coefficients. The difficulties of obtaining accurate results that are often encountered in practice are explained by analytically relating the variation of the input quantities — such as sample surface area or reverberation time — to the variation of the absorption and scattering coefficient. Special focus is set on the spatial variation of reverberation times as the primary uncertainty factor. The predicted uncertainty is successfully validated with measurements in both full-scale and small-scale reverberation chambers. Based on the uncertainty analysis, a method is developed to determine the necessary minimum number of source-receiver combinations in the sound field to ensure a specified precision of the absorption or scattering coefficient.

The second part of the thesis focuses on signal processing steps related to the measurement of angle-dependent reflection properties in the free-field. For this purpose a hemispherical microphone array is described and validated in this thesis. Improvements to the subtraction method are presented that allow to include the source and receiver directivity. Sound reflection models of different accuracy and calculation complexity are considered to deduce the surface impedance from measured reflection factors. Array processing techniques are investigated as an alternative method to obtain a source reference signal *in-situ* and to process the spatial response of the reflection measurement.

Measurements show that the array setup can be used to obtain the angle-dependent absorbing properties of samples with few source positions. The results indicate that for receivers close to the surface the simplified plane wave model should not be used as it leads to large errors, especially at low frequencies. Some uncertainty remains in the phase angle of the complex reflection factor, which is due to incomplete knowledge of the source and receiver positions. Nonetheless, relatively stable results can be obtained even for samples of finite extent. With the help of array-processing methods, the setup can also be used to determine the directional diffusion and scattering coefficient of small samples, yielding the same result as established far-field methods.

Contents

1

Introduction

Today's easy access to computing power and the development of highly advanced algorithms have made the simulation of large-scale room acoustics situations feasible and available to a broad audience. Being able to predict the room impulse response for sound propagation between sources and receivers of arbitrary composition in a given space enables entirely new approaches to the study of sound fields and the perception of a person in such a sound field. This broad area of research is not restricted to room acoustics in the traditional sense, relating primarily to concert halls and other types of auditoria. Especially in connection with a visual representation of equally high quality, virtual environments can be used for multi-disciplinary investigations relating to the fields of architecture, urban planning and psychology, among others [1, 2].

Whether the simulated sound field in virtual environments is accepted as realistic — thus leading to a feeling of immersion in the virtual scene — depends on many factors in the processing chain, the most important of which are:

- the (room) model, usually based on CAD data supplied by architects

- correct descriptions of the source and receiver, including position and orientation as well as sound power and directivity pattern

- accurate values for the boundary conditions, i.e. surface reflection characteristics

- exact sound propagation and reflection models for the entire audible frequency range, including both wave-based as well as geometrical simulation methods

- a correct reproduction of the simulated sound field by means of either headphones or loudspeakers

This thesis treats the problem of obtaining accurate values for the boundary conditions needed in room acoustics simulations. Together with the sound source properties, the reflection properties of surfaces are considered as most important to achieve a realistic sound field simulation, as has also been noted by ARETZ [3].

While the prediction of sound propagation in rooms has been well studied and can be performed with a very high accuracy, the final simulation result can only be as accurate and realistic as the input data. For the wave-based simulation methods, which are mainly used for low to medium frequencies to accomplish a correct calculation of the sound field, the required input data is the complex acoustic impedance or reflection factor of all boundary surfaces. The geometrical methods, which can be used as an approximation to the wave-based methods at higher frequencies, work with energetic quantities, i. e. the absorption and scattering coefficient.

Regarding the boundary conditions, both the absorbing as well as the scattering properties of surfaces are considered here. The study is split into two parts: in the first part, the standardized methods for measurements of the energetic absorption and scattering coefficient for random sound incidence in a reverberation chamber are analyzed with respect to their susceptibility to measurement uncertainties. The second part focuses on the free-field measurement of angle-dependent reflection properties, including both the complex reflection factor as well as the directional diffusion and scattering coefficient. Both topics will be introduced in the following paragraphs.

1.1 Uncertainty Analysis of Reverberation Chamber Measurements

Measurements in the reverberation chamber and their precision is probably one of the most researched fields of acoustics. Based on the relationship between the average room absorption and the reverberation time derived by SABINE [4], absorbing and scattering properties of samples in a diffuse sound field can be measured. The procedure is described in ISO 354 [5] and ASTM C423 [6] for the absorption coefficient and in ISO 17497-1 [7] for the scattering coefficient. Being the most common methods to obtain data of materials used in architectural acoustics, the analysis of reverberation chamber measurements is of special importance.

The dimensions of the reverberation chamber were found to be a source of uncertainty [8, 9] as was the placement of the sample [10, 11]. Often, the measured absorption coefficient exceeds the physically plausible value of one, which has been related to the edges of the sample [12, 13, 14, 15]. The violation of the theoretical assumptions of the measurement procedure concerning a diffuse sound field is a field of study related to intra-laboratory repeatability and precision [16, 17, 18, 19, 20]. The degree of diffuseness of the sound field has been suggested to be directly related to the quality of the result [21, 22]. Many studies have investigated the inter-laboratory reproducibility of reverberation chamber measurements using round-robin tests [23, 24, 25, 26, 27, 28]. Significant variations have been determined between laboratories and efforts are currently being undertaken to establish guidelines to obtain more reliable results.

The rather new method to measure the random-incidence scattering coefficient [29] has also been investigated regarding the sample size [30]. The influence of the measurement setup has been determined experimentally by CHOI and JEONG [31]. As it is possible to analytically or numerically calculate the scattering coefficient for certain surfaces [32, 33, 34], some case studies have been carried out to validate the measurement procedure, also with respect to setups in small-scale [35, 36, 37]. Good agreement has generally been found along with a relatively high measurement uncertainty.

The concept of uncertainties and their analysis is present in all major fields of engineering [38, 39]. By now, it has been established that reporting measurement results should include an indication of the uncertainty that is connected to the average result. In acoustics, the problem of measurement uncertainties has mostly been treated experimentally. Apart from the aforementioned investigations, the influence of the signal processing steps related to the determination of reverberation times has been considered by LUNDEBY et al. [40] and GUSKI and VORLÄNDER [41]. The uncertainty of absorption coefficient measurements especially at low frequencies has been treated by DÄMMIG and DEICKE [42].

In this thesis, an analytical analysis of the effect of uncertainties on the measured absorption and scattering coefficients is presented. The approach is based on the law of error propagation [43] and the analysis follows the recommendations given by the *Guide to the expression of Uncertainty in Measurement* (GUM) [44]. All of the input quantities needed to obtain the absorption or scattering coefficient are treated regarding either systematic or random deviations from the expected value. The resulting equations to predict the uncertainty are validated by measurements where possible. For the transmission loss, a similar study has been carried out by WITTSTOCK [45]. However — to the knowledge of the author — a comprehensive analysis as the one presented here has not been done concerning measurements of reflection properties. A study on the effect of the types of uncertainties investigated here on simulation results in room acoustics has been presented by VORLÄNDER [46].

1.2 Measurement of Angle-Dependent Reflection Properties

The results obtained from reverberation chamber measurements may not provide enough information. This is the case when the complex reflection factor or impedance is needed for room acoustics simulations [47, 48, 49]. A standardized method to measure complex reflection parameters based on the Kundt's Tube is ISO 10534-2 [50]. However, the method only yields the result for normal incidence. In those cases where an angle-dependent result is desired [51], measurements have to be performed in the free-field or even *in-situ*, i.e. at the location of the installed sample.

The goal of measuring reflection properties in flexible setups with arbitrary angles of sound incidence has led to a large number of investigations. A literature review can for example be found in the thesis by GEETERE [52]. The measurement setup usually involves one source and one receiver [53], though the two-microphone method has also been applied [54, 55, 56]. A setup with many microphones and the application of the spatial Fourier transform has been used by TAMURA [57] and TAMURA, ALLARD, and LAFARGE [58]. Several works consider different measurement signals and processing methods to obtain the reflection factor [59, 60, 61, 62]. Because most methods try to deduce the reflection factor based on the measured sound pressure above the boundary, a realistic model of sound reflection is needed. Some of the most commonly employed models will be discussed and compared in this thesis.

If not only the pressure but also the particle velocity — and thus the field impedance — in front of the surface can be measured, the deduction of the surface impedance can be performed more easily [63]. The measurement of the surface impedance with a combination of a pressure and velocity sensor — so-called *pu*-probes — has been promoted recently [63, 64]. It has been found that the knowledge of the exact distance of the sensor from the surface is crucial in obtaining reliable results [65]. In [66] it has been shown that even reflections at the sensor can have a detrimental effect on the result. The aspect of an influence of the measurement setup on the sound field has also been considered in this thesis.

Concerning the scattering properties, a free-field measurement is far more complicated in comparison to the reverberation chamber method as the sound field has to be sampled with high spatial resolution [67]. First attempts to obtain scattering coefficients in the free-field have been made by SCHMICH and BROUSSE [68]. Due to its definition, the diffusion coefficient may be easier to measure in the free-field and an approach related to that by TAMURA [57] has been employed by KLEINER, GUSTAFSSON, and BACKMAN [69].

The methodology is extended to spherical instead of planar arrays in this thesis.

In the second part of this thesis, a hemispherical arrangement of microphones is presented that is used to measure sound reflection properties in the free-field. In contrast to the first part, angle-dependent and complex reflection properties are considered that can e. g. be used in wave-based simulation methods. The measurement setup, which consists of a sequential array with 24 microphones, is described in detail and the signal processing steps to obtain the desired reflection properties are discussed. The use of an array allows to consider spatial filtering methods and it is investigated whether these methods can be successfully applied to the problem at hand, especially regarding the scattering properties. Results for different absorbers and a scattering surface are presented and discussed.

2

Fundamentals

This chapter will provide the basic definitions and equations that are related to the propagation of sound and its reflection at extended surfaces, including sound scattering. Additionally, a brief introduction into the topic of statistical room acoustics will be given as a prerequisite for the standardized measurement methods. The latter will be used for the uncertainty analysis, of which the fundamentals are also briefly introduced here. As an advanced signal processing method, the transformation of data into the Spherical Wave Spectrum will be covered as well.

2.1 Sound Propagation in Free Space

Before dealing with the reflection of sound at boundaries, it is first necessary to describe the propagation of sound in free space and the elementary wave types encountered.

2.1.1 Wave Equation

The basis for the description of sound propagation in gases or fluids — in the context of this work especially in air — are the sound field equations. The medium in this context is assumed to be an ideal gas at rest, i.e. there is no flow of the medium. The sound field equations describe the relationship between the two acoustic field quantities: the vector of particle velocity $\vec{v}$ and the scalar sound pressure p. The order of magnitude of the sound pressure is assumed to be much lower than the static pressure p_0. The sound field equations in three dimensions are then given by

$$\operatorname{grad} \underline{p} = -\rho_0 \frac{\partial \underline{\vec{v}}}{\partial t} \,, \tag{2.1}$$

$$\operatorname{div} \underline{\vec{v}} = -\frac{1}{\rho_0 \, c^2} \frac{\partial \underline{p}}{\partial t} \,, \tag{2.2}$$

where c and ρ_0 are the speed of sound and the density, respectively, with typical values for air of $344 \, \mathrm{m/s}$ and $1.2 \, \mathrm{kg/m^3}$ at an ambient temperature of $20 \, °\mathrm{C}$.

Eq. (2.1) is based on the conservation of momentum and Eq. (2.2) on the conservation of mass. The complete derivation can be found in the standard text books by e. g. KUTTRUFF [70] or BERANEK [71]. Eq. (2.1) shows that the direction of the pressure gradient and hence the propagation direction of the wave is the same as for the particle velocity. This means that acoustic pressure waves in fluids are longitudinal waves.

Applying the divergence operator to Eq. (2.1) and partially differentiating Eq. (2.2) with respect to the time t leads to the three-dimensional wave equation

$$\Delta \underline{p}(\vec{r}, t) = \frac{1}{c^2} \frac{\partial^2 \underline{p}(\vec{r}, t)}{\partial t^2} \,, \tag{2.3}$$

where $\Delta = \mathrm{div} \, \mathrm{grad}$ is the Laplace operator and the vector $\vec{r}$ defines the position of the observer. In Cartesian coordinates, the Laplace operator expands to

$$\Delta \underline{p} = \frac{\partial^2 \underline{p}}{\partial x^2} + \frac{\partial^2 \underline{p}}{\partial y^2} + \frac{\partial^2 \underline{p}}{\partial z^2} \tag{2.4}$$

whereas in spherical coordinates it is defined as

$$\Delta \underline{p} = \frac{\partial^2 \underline{p}}{\partial r^2} + \frac{2}{r} \frac{\partial \underline{p}}{\partial r} + \frac{1}{r^2 \sin(\theta)} \frac{\partial}{\partial \theta} \left(\sin(\theta) \frac{\partial \underline{p}}{\partial \theta} \right) + \frac{1}{r^2 \sin^2(\theta)} \frac{\partial^2 \underline{p}}{\partial \varphi^2} \,, \tag{2.5}$$

where the following coordinate transform pairs were used for the radius r, polar angle θ and azimuth angle φ:

$$r = \sqrt{x^2 + y^2 + z^2} \,, \tag{2.6}$$

$$\theta = \arccos\left(\frac{z}{r}\right) \,, \tag{2.7}$$

$$\phi = \arctan\left(\frac{y}{x}\right) \,, \tag{2.8}$$

with $r \in \mathbb{R}_{\geq 0}$, $\theta \in [0, \pi]$ and $\phi \in [0, 2\pi)$.

In the following, harmonic signals (with the angular frequency $\omega = 2\pi f$) of the form $\underline{p}(t) = \hat{p} \, \mathrm{e}^{\mathrm{j}\omega t}$ will be assumed, with the imaginary unit defined as $\mathrm{j} = \sqrt{-1}$. The wave equation in Eq. (2.3) can then be transformed into the homogeneous Helmholtz equation by replacing the differentiation with respect to time by a multiplication with $\mathrm{j}\omega$:

$$\Delta \underline{p}(\vec{r}, t) + k^2 \underline{p}(\vec{r}, t) = 0 \,, \tag{2.9}$$

where $k = \frac{\omega}{c} = \frac{2\pi}{\lambda}$ is the angular wavenumber and λ the wavelength.

2.1.2 Plane Waves

The most simple — and yet most commonly applied — model for propagating waves is that of plane waves, in which the field quantities only vary along the direction of propagation while on a plane perpendicular to that direction the quantities are constant.

Any twice-differentiable function of the form

$$\underline{p}(\vec{r}, t) = \underline{f}(\vec{n}\vec{r} - ct) \tag{2.10}$$

with $||\vec{n}|| = 1$ is a solution of Eq. (2.9). It describes planes of constant sound pressure (and thereby of constant particle velocity) traveling in normal direction $\vec{n}$ at the speed of sound c. The approach

$$\underline{f}(k, x) = \hat{p}\, e^{-jkx} \tag{2.11}$$

yields the expression for a harmonic plane wave

$$
\begin{aligned}
\underline{p}(k, \vec{r}, t) &= \hat{p}\, e^{-jk(\vec{n}\vec{r} - ct)} \\
&= \hat{p}\, e^{j(\omega t - \vec{k}\vec{r})}
\end{aligned} \tag{2.12}
$$

where the wavenumber vector $\vec{k}$ is defined as $\vec{k} = k\,\vec{n}$.

It should be noted at this point that the plane wave is a rather theoretical construct and is almost never encountered in real life.[1] Plane waves simply exist: they have no origin, i.e. there is no source, and there is no loss of energy during propagation. This means that there is no reduction of level with increasing propagation distance, a fact that is obviously not encountered in real situations. These properties arise from solving the homogeneous Helmholtz equation. As will be seen in the next subsection, a more physically plausible solution can be found by including a source term in Eq. (2.9).

An important quantity of the propagation medium is the so-called *field impedance*, i.e. the ratio of sound pressure and particle velocity $\underline{v}_n$ (in the direction of wave propagation) for a wave in free-space

$$\underline{Z}_f = \frac{\underline{p}}{\underline{v}_n}, \tag{2.13}$$

---

[1] Only guided waves in tube-like volumes approximately show the form of plane waves.

which, in the case of a plane wave, is equal to the *characteristic impedance* of the medium

$$Z_0 = \underline{Z}_f\big|_{\text{plane wave}} = \rho_0\, c \qquad (2.14)$$

with a typical value of approximately $Z_0 = 412.5\,{}^{\text{kg}}\!/_{(\text{s}\,\text{m}^2)}$ for air at $20\,^\circ\text{C}$.

Eq. (2.14) shows that the relationship between sound pressure and particle velocity in a plane wave does not depend on frequency and that the two field quantities are always in phase.

2.1.3 Spherical Waves

While for the plane wave the idea of a sound source was not included, the spherical wave is a more useful concept as it describes the sound field due to a source at a specific position in free space, which emits a sound wave whose wavelength is much bigger than the extent of the source. In analogy to the plane wave, the surfaces of constant sound pressure and particle velocity for the spherical wave are concentric spheres centered at the position of the source.

Including a source term on the right-hand side of Eq. (2.9) leads to the inhomogeneous Helmholtz equation. The approach (compare Eq. (2.10))

$$\underline{p}(r,t) = \frac{1}{r}\underline{f}(r - ct) \qquad (2.15)$$

then gives a solution to the inhomogeneous Helmholtz equation in spherical coordinates, where the spatial variation of the sound pressure only depends on the distance r between source and receiver.

A source which emits waves in the way described here is called a point source and it is characterized by its volume velocity $Q(t)$, i. e. the volume expelled per time unit. Without further derivation (again, found e. g. in [70]) Eq. (2.15) becomes

$$\underline{p}(r,t) = \frac{\rho_0}{4\pi r}\frac{\partial}{\partial t}\underline{Q}\!\left(t - \frac{r}{c}\right). \qquad (2.16)$$

Again, assuming harmonic excitation $\underline{Q}(t,\omega) = \hat{Q}\,\mathrm{e}^{\mathrm{j}\omega t}$ yields

$$\underline{p}(r,t,\omega) = \frac{\mathrm{j}\omega\rho_0\hat{Q}}{4\pi r}\mathrm{e}^{\mathrm{j}(\omega t - kr)} = \mathrm{j}\omega\rho_0\hat{Q}\cdot\underline{G}(k,r,t)\,, \qquad (2.17)$$

where the so-called *Green's function*, which describes the acoustical transfer function between a source and a receiver in three-dimensional free space, has been introduced:

$$\underline{G}(k,r,t) = \frac{1}{4\pi r}\mathrm{e}^{\mathrm{j}(\omega t - kr)}\,. \qquad (2.18)$$

For a better readability, the time-dependent factor $e^{j\omega t}$ will be omitted in all following equations. Also, the dependency on the angular frequency ω (or the wavenumber k) will sometimes be suppressed in the notation but is considered implicitly.

The field impedance in the case of the spherical wave is

$$\underline{Z}_f(k,r) = \frac{Z_0}{1 + \frac{1}{jkr}}, \qquad (2.19)$$

which shows that, contrary to the plane wave, in the spherical wave the two field quantities are not necessarily in phase and that the ratio is frequency-dependent. Figure 2.1 shows a plot of the modulus and phase of the specific field impedance as a function of r/λ for plane and spherical waves.

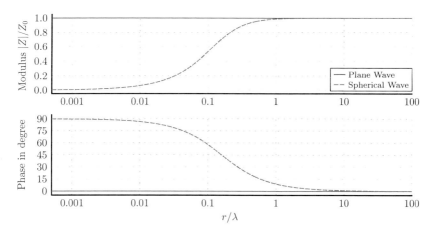

Figure 2.1: Modulus and phase of the specific field impedance as a function of r/λ for the plane wave and spherical wave

The graph shows that for $r/\lambda > 1$ the phase offset between pressure and particle velocity vanishes and the characteristic impedance of the plane wave is reached. This means that there is a region (whose extent is frequency-dependent), the so-called *near-field*, where the complex relation between the field quantities has to be taken into account. On the other hand, at distances several wavelengths away from the source, hence called the *far-field*, the wavefronts can be treated as stemming from plane waves.

2.1.4 Influence of Atmospheric Conditions

Density and speed of sound The properties of the medium in which sound waves propagate, i.e. air in this context, obviously affect the propagation. Here, especially the temperature and relative humidity are important as they have an influence on the density ρ_0 and the speed of sound c and thus on the specific impedance $Z_0 = \rho_0 c$. According to ISO 2533 [72], which describes the properties of the standard atmosphere under the assumption of air as an ideal gas, the density and speed of sound, respectively, depend on the temperature as

$$\rho_0(\Theta) = \frac{p_{\text{stat}}}{R\Theta}\,, \tag{2.20}$$

and

$$c(\Theta) = \sqrt{\kappa R\Theta} = \sqrt{\kappa \cdot \frac{p_{\text{stat}}}{\rho_0(\Theta)}}\,, \tag{2.21}$$

where Θ is the absolute temperature in Kelvin, p_{stat} is the static air pressure, $\kappa = 1.4$ the adiabatic exponent and $R = 287.05287\,\text{J}/(\text{K kg})$ the specific gas constant of air. The equation for the speed of sound in air can be linearized to yield the commonly used relationship

$$c(\Delta\Theta) = (331.4 + 0.6\,^1/\!^\circ\text{C} \cdot \Delta\Theta)\,\text{m/s}\,, \tag{2.22}$$

with the temperature difference (i.e. the Celsius scale) $\Delta\Theta = (\Theta - \Theta_0)$ where $\Theta_0 = 273.15\,\text{K}$ is the reference temperature (as the melting point of ice). The linearization in Eq. (2.22) leads to a relative error in the speed of sound of at most $0.2\,\%$ in relation to Eq. (2.21) in the temperature range of $0\,^\circ\text{C} \le \Delta\Theta \le 40\,^\circ\text{C}$.

In Eq. (2.20)–Eq. (2.22) the effect of relative humidity on the density and speed of sound has been neglected. Especially concerning practical applications the influence is negligible and leads to relative errors of less than $2\,\%$ for the density and less than $1\,\%$ for the speed of sound, all calculated for relative humidities in the typical range of $30\,\% \le \phi \le 90\,\%$.

Air attenuation The atmospheric conditions do not only affect the wave propagation in terms of the speed of sound and the specific wave impedance, but the air as a medium also introduces an amplitude reduction that depends on the propagated distance Δr and the so-called *air attenuation coefficient* m (in $^1/\text{m}$) [73].

According to ISO 9613-1 [74], which gives the equations to calculate the air attenuation coefficient (allegedly with an estimated uncertainty of $\pm 10\%$), the sound pressure is reduced by air attenuation as follows[2]:

$$\frac{\underline{p}(r + \Delta r)}{\underline{p}(r)} = \mathrm{e}^{-\frac{m}{2}\Delta r} . \tag{2.23}$$

Air attenuation is roughly proportional to frequency and is thus important especially at high frequencies. It furthermore depends on the relative humidity and temperature and for this reason the atmospheric conditions should be recorded carefully during measurements. This can be confirmed with the data graphed in Figure 2.2a and Figure 2.2b, which show the air attenuation coefficient in dB/km calculated according to ISO 9613-1 [74] as a function of frequency and for different values of relative humidity and temperature, respectively. In each figure, the solid purple curve corresponds to the typical average values of $\Delta\Theta = 20\,^\circ\mathrm{C}$ and $\phi = 60\,\%$, which will be considered as the reference values in all further investigations.

The concept of air attenuation can be implemented in the propagation terms of Eq. (2.12) and Eq. (2.17) by considering the wavenumber to be complex (see [70, Section 4.3]):

$$\underline{k} = k - \mathrm{j}\frac{m}{2} = \frac{\omega}{c} - \mathrm{j}\frac{m}{2} . \tag{2.24}$$

This has the advantage that the previous equations relating to wave propagation do not have to be modified. In all following equations air attenuation is not explicitly considered but as mentioned it can be integrated with ease.

---

[2]ISO 9613-1 denotes the amplitude attenuation factor by α, whereas here and in general in room acoustics the energy attenuation coefficient m is used, hence the factor $1/2$.

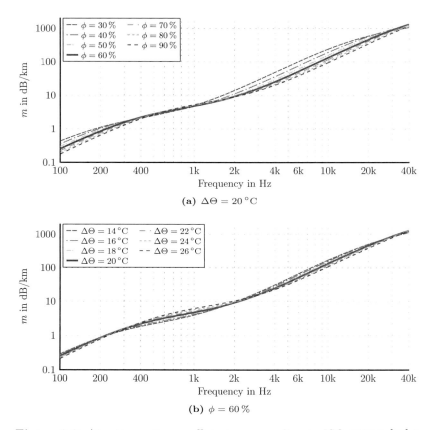

(a) $\Delta\Theta = 20\,°\mathrm{C}$

(b) $\phi = 60\,\%$

Figure 2.2: Air attenuation coefficient m according to ISO 9613-1 [74] in dB/km as a function of frequency and for different values of the temperature $\Delta\theta$ and relative humidity ϕ

2.2 Spherical Wave Spectrum

Many problems in acoustics, especially those related to sound radiation from point-like sources, are easily formulated in a spherical coordinate system and the solution of the wave equation in Eq. (2.3) in spherical coordinates is then applied. This leads to the use of the so-called *Spherical Harmonics* (SH), which describe an orthonormal set of (complex) functions depending on the polar angle θ and azimuth angle φ.

By the use of Spherical Harmonics, the *Spherical Harmonics Transform* (SHT) of data obtained on a spherical grid of receivers into the *Spherical Wave Spectrum* (SWS) can be performed. The analysis of data in the SH domain can be helpful for many radiation problems. In this thesis, the SH analysis is only used as a tool and hence the theory will not be provided in much detail. The reader is referred to existing literature on the topic for further information, e.g. by WILLIAMS [75] and ZOTTER [76]. However, it has to be noted that due to different sign and normalization conventions, the formulas may differ from those in the standard literature. In that respect, this section is more closely related to the work by POLLOW [77].

2.2.1 Spherical Harmonics

The complex-valued Spherical Harmonics function $\underline{Y}_n^m$ of integer-valued order n and degree m describing the angular dependency of the sound field is defined by

$$\underline{Y}_n^m(\theta, \varphi) = \sqrt{\frac{(2n+1)}{4\pi} \frac{(n-m)!}{(n+m)!}} \cdot P_n^m(\cos(\theta))\, \mathrm{e}^{\mathrm{j}m\varphi}, \qquad (2.25)$$

where $P_n^m(x)$ is the associated Legendre function. The set of SH functions can then be used to apply the SHT to any kind of square-integrable function $\underline{p}(\theta, \varphi)$, representing for example pressure data, on the sphere:

$$\underline{p}_{nm} = \int\limits_0^{2\pi} \int\limits_0^{\pi} \underline{p}(\theta, \varphi) \underline{Y}_n^{m^*}(\theta, \varphi) \sin(\theta)\, \mathrm{d}\theta\, \mathrm{d}\varphi, \qquad (2.26)$$

where $\underline{Y}^*$ denotes the complex conjugate of $\underline{Y}$. If instead the SH coefficients $\underline{p}_{nm}$ are known, the *Inverse Spherical Harmonics Transform* (ISHT) can be performed to obtain the result in the spatial domain

$$\underline{p}(\theta, \varphi) = \sum_{n=0}^{\infty} \sum_{m=-n}^{n} \underline{p}_{nm}\, \underline{Y}_n^m(\theta, \varphi). \qquad (2.27)$$

15

The SHT and ISHT are exact only as long as the integration in Eq. (2.26) can be accurately performed or equivalently if the summation in Eq. (2.27) is carried out for orders up to infinity. In practical situations the resolution of the spherical grid of N_p receivers limits the spatial resolution and hence the maximum order $N_{\max}$ that can be used for the SHT. If the SH functions $\underline{Y}_n^m$ are stacked into a matrix $\mathbf{Y}$ of size $\left[N_p \times (N_{\max} + 1)^2\right]$ and the coefficients $\underline{p}_{nm}$ into a vector $\mathbf{p_{nm}}$ of size $\left[(N_{\max} + 1)^2 \times 1\right]$ the ISHT reduces to a simple matrix-vector product

$$\mathbf{p} = \mathbf{Y}\mathbf{p_{nm}}\,, \tag{2.28}$$

where $\mathbf{p}$ is of size $[N_p \times 1]$. The SHT involves the inverse of the matrix of SH functions:

$$\mathbf{p_{nm}} = \mathbf{Y}^{(-1)}\mathbf{p}\,, \tag{2.29}$$

where the operator $\mathbf{Y}^{(-1)}$ suggests that the inverse of the matrix $\mathbf{Y}$ may not always exist and hence special inversion methods have to be used. This is strongly related to the spatial sampling used for the receiver grid so that no general solution can be presented. Usually a pseudo (or least-squares) inversion or a regularized inversion is employed. Further information can for example be found in [78].

2.2.2 Plane and Spherical Waves

In analogy to the equation in Cartesian coordinates (Eq. (2.12)), the sound pressure of a plane wave impinging from an angular direction (θ_0, φ_0) onto a spherical array of radius r can be formulated in the SH domain by

$$\underline{p}_{nm}(k, r) = 4\pi\, \mathrm{j}^n\, j_n(kr)\, \underline{Y}_n^{m*}(\theta_0, \varphi_0)\,, \tag{2.30}$$

where $j_n(x)$ is the spherical Bessel function of order n. In addition to the angular dependency represented by the SH functions, the radial Bessel term is needed to describe propagating waves.

For the case of a point source at a location (R, θ_0, φ_0) in spherical coordinates, the SH coefficients can be calculated depending on the relation between the source radius R and the array radius r as

$$\underline{p}_{nm}(k, r) = -\mathrm{j}k\, \underline{Y}_n^{m*}(\theta_0, \varphi_0) \cdot \begin{cases} j_n(kr)\, h_n^{(2)}(kR) & R > r\,, \\ h_n^{(2)}(kr)\, j_n(kR) & R < r\,, \end{cases} \tag{2.31}$$

where $h_n^{(2)}$ is the spherical Hankel function of the second kind of order n. For $R \to \infty$ the upper branch in Eq. (2.31) simplifies to the plane wave solution in Eq. (2.30) (see [76, Section 2.4.2]).

In order to gain an insight into the general behavior of the functions describing wave propagation in the SH domain, Figure 2.3a and Figure 2.3b show the spherical Bessel and Hankel function, respectively, for various orders n as a function of kr. It can be seen that the spherical Bessel function has many zero crossings and that it decreases towards low frequencies proportionally to $(kr)^n$. In comparison, the spherical Hankel function is a smooth function but increases towards low frequencies proportionally to $(kr)^{n+1}$ (see also [75, Section 6.4]).

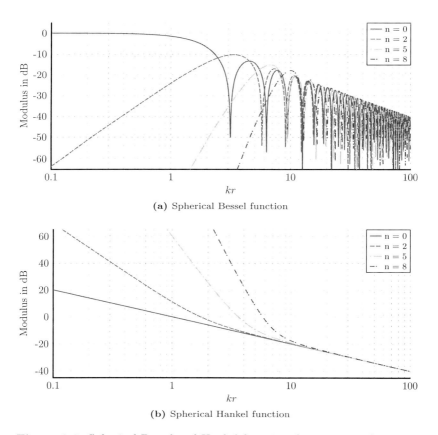

(a) Spherical Bessel function

(b) Spherical Hankel function

Figure 2.3: Spherical Bessel and Hankel functions for various orders n as a function of kr

In the following, one of the tasks is to distinguish between incoming and outgoing waves due to sources outside and inside the receiver array, respectively. Since the distance of the source is usually not of interest, the coefficients can be summarized as

$$\underline{A}_{nm} = -jk\,\underline{Y}_n^{m^*}(\theta_A, \varphi_A)\,h_n^{(2)}(kR_A)\,, \tag{2.32}$$

for incoming waves and

$$\underline{B}_{nm} = -jk\,\underline{Y}_n^{m^*}(\theta_B, \varphi_B)\,j_n(kR_B)\,, \tag{2.33}$$

for outgoing waves. The sound pressure coefficients can then be simply written as

$$\underline{p}_{nm}(k, r) = \underline{A}_{nm}\,j_n(kr) + \underline{B}_{nm}\,h_n^{(2)}(kr)\,. \tag{2.34}$$

The presented equations can of course be generalized to any arbitrary number of sources inside and outside of the array by simply summing the contributions from each individual source.

2.2.3 Scattering Near-Field Holography

If the coefficients $\underline{p}_{nm}$ have been determined, Eq. (2.34) presents one equation with two unknowns, $\underline{A}_{nm}$ and $\underline{B}_{nm}$, which cannot be uniquely solved. This problem can be overcome by obtaining $\underline{p}_{nm}$ on two spherical shells of different radii r_1 and r_2. Figure 2.4 presents a schematic of this situation with two sources, one each inside and outside of the receiver arrays.

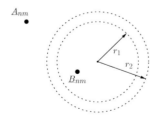

Figure 2.4: Exemplary setup for scattering near-field holography with a source outside (A_{nm}) and inside (B_{nm}) of the arrays of radii r_1 and r_2

In that case, the following system of equations is obtained:

$$\underline{p}_{nm}(k, r_1) = \underline{A}_{nm}\, j_n(kr_1) + \underline{B}_{nm}\, h_n^{(2)}(kr_1)\,,$$
$$\underline{p}_{nm}(k, r_2) = \underline{A}_{nm}\, j_n(kr_2) + \underline{B}_{nm}\, h_n^{(2)}(kr_2)\,.$$

The solution involves the matrix determinant operator

$$\begin{vmatrix} a & b \\ c & d \end{vmatrix} = a \cdot d - b \cdot c\,, \tag{2.35}$$

using which the coefficients can be calculated by

$$\underline{A}_{nm} = \frac{\begin{vmatrix} \underline{p}_{nm}(k, r_1) & h_n^{(2)}(k, r_1) \\ \underline{p}_{nm}(k, r_2) & h_n^{(2)}(k, r_2) \end{vmatrix}}{\Delta}\,, \tag{2.36}$$

$$\underline{B}_{nm} = \frac{\begin{vmatrix} j_n(k, r_1) & \underline{p}_{nm}(k, r_1) \\ j_n(k, r_2) & \underline{p}_{nm}(k, r_2) \end{vmatrix}}{\Delta}\,, \tag{2.37}$$

with

$$\Delta = \begin{vmatrix} j_n(k, r_1) & h_n^{(2)}(k, r_1) \\ j_n(k, r_2) & h_n^{(2)}(k, r_2) \end{vmatrix}\,. \tag{2.38}$$

The prerequisite for these equations to be correct is that the region $r_1 < r < r_2$ is free of sources. This method is known as *Scattering Near-Field Holography*, which has been applied, for example, by WEINREICH and ARNOLD [79] to determine the sound radiation from violins.

Obviously the solution becomes instable for zero (or almost-zero) values of Δ. The first of such values is encountered when $k(r_2 - r_1) = \pi$; in other words, the frequency given by

$$f_{\text{alias,radial}} = \frac{c}{2 \cdot (r_2 - r_1)} \tag{2.39}$$

determines an upper limit for the usable frequency range for a given array setup due to radial aliasing. Figure 2.5 shows the behavior of (the level of) Δ according to Eq. (2.38) for various orders as a function of $k(r_2 - r_1)$. As can be seen, the aliasing frequency does not depend on the order n but is only determined by the geometrical setup.

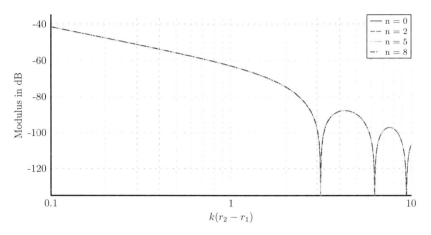

Figure 2.5: Δ (according to Eq. (2.38)) in dB for various orders as a function of $k(r_2 - r_1)$

2.2.4 Spherical Beamforming

One of the most commonly applied spatial filtering methods related to microphone arrays of arbitrary shapes is called *Beamforming* and the most simple and yet commonly used algorithm is the *Delay-and-Sum Beamformer* [80]. In analogy, beamforming filters can also be designed using the SH base functions and appropriate radial filters.

In fact, it was shown by RAFAELY [81] that the Delay-and-Sum algorithm can be directly transferred to the SH domain to work on the pressure coefficients $\underline{p}_{nm}$, yielding the output signal

$$\underline{y}(k,r) = \frac{1}{4\pi} \sum_{n=0}^{N_{\max}} \sum_{m=-n}^{n} \underline{w}^*_{nm}(k,r)\, \underline{p}_{nm}(k,r)\,, \qquad (2.40)$$

where the beamforming weights $\underline{w}_{nm}$ for the look (or *steering*) direction (θ_l, φ_l) are given by:

$$\underline{w}_{nm}(k,r) = \underline{b}_{nm}(k,r)\, \underline{Y}_n^{m^*}(\theta_l, \varphi_l)\,. \qquad (2.41)$$

The weighting coefficients $\underline{b}_{nm}$, which predict the phase distribution across the array microphones, are chosen according to the employed wave model. Again, similar to the traditional beamforming algorithms [82], either far-field (i. e. plane-wave) beamforming can be employed, leading to (compare Eq. (2.30))

$$\underline{b}_{nm}(k,r) = 4\pi\, \mathrm{j}^n\, j_n(kr)\,, \qquad (2.42)$$

or near-field beamforming assuming spherical waves is used, which gives the coefficients (compare Eq. (2.31))

$$\underline{b}_{nm}(k,r) = -\mathrm{j}k\, j_n(kr)\, h_n^{(2)}(kR_l)\,, \qquad (2.43)$$

for $R_l > r$, i. e. scanning positions outside of the array. In the case of near-field beamforming, a scanning distance R_l is a parameter in addition to the looking direction.

Since the source position is known in the application described in this work, near-field beamforming will be applied because — in the ideal case — it allows to determine the absolute source amplitude with the correct phase. In that case the beamforming output has to be scaled to obtain the correct source amplitude:

$$\underline{y}'(k,r) = \underline{y}(k,r) \cdot (4\pi R_l)^2\,, \qquad (2.44)$$

because

$$\underline{p}_{nm}, \underline{w}_{nm} \propto \frac{1}{4\,\pi R_l}\,. \qquad (2.45)$$

In the application in Chapter 4, the level-corrected beamformer according to Eq. (2.44) using the near-field coefficients in Eq. (2.43) will be used.

2.2.5 Spherical Harmonics on Incomplete Spheres

It is not always possible to construct arrays that span the entire sphere, as is the prerequisite for the SHT in Eq. (2.26) to be exact. Basically, three options exist for the transformation of data on incomplete spheres:

1. **Regularized matrix inversion**: For relatively small gaps, regularization can be employed as shown by POLLOW et al. [83] to obtain a solution of minimal energy in the missing region. This solution however can become numerically unstable as the size of the gap increases and will hence not be considered here.

2. **Selection of SH functions**: If some kind of symmetry or periodicity is present in the setup this can be exploited by reducing the set of base functions to those exhibiting the same symmetry or periodicity, which would then still represent an orthonormal basis. An example is a hemispherical setup above a perfectly reflecting ground, so that

symmetry with respect to the reflecting plane is given. This example and others have been discussed by POMBERGER and ZOTTER [84]. However it should be noted that as soon as the symmetry/periodicity is not guaranteed, a transformation with such a subset of base functions practically enforces symmetry/periodicity and may thus lead to false results.

3. **Orthonormal base functions on the bounded domain**: The most general but also potentially most unstable approach consists of obtaining new orthonormal base functions $\underline{\hat{Y}}$ on the bounded domain. This is preferably done in such a way that the new SH coefficients $\underline{\hat{p}}_{nm}$ can be converted into the ones relating to the full sphere in order to be able to use the methods described in the previous paragraphs. In the following, this method will be described in more detail.

In all following derivations, it is assumed that the SHT is performed using weighted quadrature, i. e.

$$\mathbf{p_{nm}} = (\mathbf{WY})^{H}\,\mathbf{p} = \mathbf{Y}^{H}\mathbf{W}\,\mathbf{p}\,, \qquad (2.46)$$

where $\mathbf{Y}^{H}$ denotes the conjugate (or *Hermitian*) transpose of $\mathbf{Y}$ and $\mathbf{W} = \mathrm{diag}\{\mathbf{w}\}$ is a diagonal matrix of real weights ensuring orthonormality of the base functions. The weights are related to the surface sampled by each receiver point and hence have to be calculated based on the sampling strategy. Eq. (2.46) shows that the matrix of base functions actually does not have to be inverted, which has an enormous numerical advantage regarding stability. This is one of the reasons why the receiver arrays used in this thesis will always follow Gauss-Legendre quadrature sampling schemes, where the sensor positions and weights are calculated based on the zeros of Legendre polynomials [78, 85]).

The approach of obtaining relations between the original matrix of base functions $\mathbf{Y}$ and the new base functions on the bounded domain $\hat{\mathbf{Y}}$ described by PAIL, PLANK, and SCHUH [86] is based on the so-called *Gram* matrix $\mathbf{G}$, which can be calculated by analytical or numerical integration (here carried out numerically using the matrix product) of the original base functions over the bounded domain $\hat{S}$.

For the discrete SHT as in Eq. (2.46), the Gram matrix is given by[3]

$$\mathbf{G} = \mathbf{Y}_{\hat{S}}^{H} \, \mathbf{W}_{\hat{S}} \mathbf{Y}_{\hat{S}} \,, \tag{2.47}$$

where $\mathbf{W}_{\hat{S}}$ is the diagonal matrix of weights corresponding to the sampling points on the bounded domain. For the trivial case that $\hat{S}$ spans the entire sphere, $\mathbf{G} = \mathbf{I}$, with $\mathbf{I}$ being the identity matrix.

The connection between the original and the new base functions is based on the reconstruction matrix $\mathbf{R}$, related to the Gram matrix as follows:

$$\mathbf{G} = \mathbf{Y}_{\hat{S}}^{H} \, \mathbf{W}_{\hat{S}} \mathbf{Y}_{\hat{S}} = \mathbf{R}^{H} \, \mathbf{R} \,. \tag{2.48}$$

To obtain $\mathbf{R}$, first the (symmetric) Gram matrix is transformed using singular value decomposition (SVD) into the matrix of eigenvectors $\mathbf{V}$ and the vector of eigenvalues $\mathbf{s}$

$$\mathbf{G} = \mathbf{V} \operatorname{diag}\{\mathbf{s}\} \mathbf{V}^{H} \,. \tag{2.49}$$

For a truncated SVD the system can be reduced in dimension by neglecting the eigenvectors corresponding to eigenvalues below a certain threshold relative to the maximum value. The truncation can help to produce a result that is numerically more stable as the condition number is reduced.

The result in Eq. (2.49) can then be used in connection with QR-decomposition to obtain the relation

$$\mathbf{V} \operatorname{diag}\{\mathbf{s}\} \mathbf{V}^{H} = \mathbf{V} \operatorname{diag}\{\mathbf{s}^{1/2}\} \operatorname{diag}\{\mathbf{s}^{1/2}\} \mathbf{V}^{H} = \mathbf{R}^{H} \mathbf{Q}^{H} \mathbf{Q} \mathbf{R} \,, \tag{2.50}$$

finally yielding the equation for the upper-triangular reconstruction matrix

$$\mathbf{R} = \mathbf{Q}^{H} \operatorname{diag}\{\mathbf{s}^{1/2}\} \mathbf{V}^{H} \,, \tag{2.51}$$

where the orthogonality property $\mathbf{Q}^{H} \mathbf{Q} = \mathbf{I}$ has been exploited.

With the reconstruction matrix and its (pseudo-)inverse $\mathbf{R}^{+}$, the connection is established regarding the base functions

$$\hat{\mathbf{Y}} = \mathbf{Y}_{\hat{S}} \, \mathbf{R}^{+} \,, \tag{2.52}$$

and the SH coefficients

$$\hat{\mathbf{p}}_{\mathbf{nm}} = \mathbf{R} \, \mathbf{p}_{\mathbf{nm}} \,. \tag{2.53}$$

The effect of the reconstruction matrix is that the base functions $\hat{\mathbf{Y}}$ will be linear combinations of the base functions $\mathbf{Y}$, possibly of different orders

---

[3] Due to different definitions and the fact that only real functions are considered in [86], slightly different matrices and equations regarding the Gram matrix $\mathbf{G}$ and the reconstruction matrix $\mathbf{R}$ are obtained here.

and degrees and hence there may no longer be a clear connection of order and degree for the newly found base functions. This is why the inverse transformation of Eq. (2.53) is necessary before applying radial filters as defined in Section 2.2.4 and Section 2.2.3.

In Chapter 4, the applicability of the SHT on incomplete spheres to the measurement of surface reflection properties will be investigated. This will also include the investigation regarding the use and stability of beamforming and scattering near-field holography on bounded domains.

2.3 Sound Reflection at an Extended Planar Surface

After having described how sound propagates in free space, the reflection of a sound wave at a single planar surface (or more general at the planar boundary between two different media) is treated. In the theoretical treatment it will be assumed that the boundary is of large extent in comparison to the wavelength, i. e. effects of edge diffraction will not be considered. Another condition for the theoretical descriptions is that the medium below the boundary is homogeneous, so that it can be described by its material parameters. For the sake of simplicity, first the concept of plane waves will be used and later more complex scenarios such as the reflection at locally and laterally reacting materials (Section 2.3.3) and of spherical waves (Section 2.3.4) will be discussed.

This section is largely based on the excellent introduction to this topic by MECHEL [87].

2.3.1 General Problem Formulation

As an introduction to the problem of sound reflection at the boundary between two media with different acoustic properties, the schematic in Figure 2.6 is considered.

Here, and in the following descriptions, the lossless medium in the upper half-space ($z > 0$) is air, characterized by the real-valued characteristic impedance[4] $Z_0 = \rho_0 c_0$ and propagation constant $k_0 = \frac{\omega}{c_0}$. The medium in the lower half-space ($z < 0$), characterized by the complex characteristic impedance $\underline{Z}_1 = \underline{\rho}_1 \frac{\omega}{\underline{k}_1}$ with the complex propagation constant $\underline{k}_1$, could be a

---

[4]The extension to include losses due to air attenuation has been covered in Section 2.1.4 and can also be applied here without loss of generality. The term for the characteristic impedance then changes to $\underline{Z}_0 = \rho_0 \frac{\omega}{\underline{k}_0}$.

fluid with different acoustic properties or — with reference to the application of the model in this thesis — it could be an acoustic absorber.

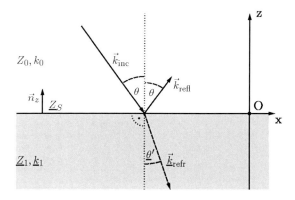

Figure 2.6: Schematic for plane wave sound reflection and refraction at the boundary between two media with different acoustic properties

An incoming wave, originating in the upper half-space, is then reflected at the boundary ($z = 0$) and possibly refracted partially into the lower half-space. The incoming plane wave is characterized by the propagation vector $\vec{k}_{\text{inc}}$ and impinges on the boundary at an angle θ measured from the normal $\vec{n}_z$.

The reflected wave with the propagation vector $\vec{k}_{\text{refl}}$ follows the law of reflection, which states that the angle of reflection is the same as the angle of incidence. For the wave reflected at the boundary this means that its direction does not depend on the medium in the lower half-space; its amplitude, however, does. This change in amplitude is usually represented by the so-called *plane wave reflection factor* $\underline{R}$, which is multiplied by the amplitude of the incoming wave $\underline{\hat{p}}_{\text{inc}}$ to obtain the amplitude of the reflected wave $\underline{\hat{p}}_{\text{refl}}$, i.e. (compare Eq. (2.12))

$$\underline{\hat{p}}_{\text{refl}} = \underline{R} \cdot \underline{\hat{p}}_{\text{inc}} . \tag{2.54}$$

However, the refracted wave, i.e. the wave that is transmitted into the medium in the lower half-space, may travel in a different direction depending on the medium. Demanding continuity of pressure and normal particle velocity at the boundary leads to the law of refraction (also called *Snell's law*) for lossy media:

$$k_0 \sin(\theta) = \underline{k}_1 \sin(\underline{\theta'}) , \tag{2.55}$$

where the angle of refraction $\underline{\theta}'$ can become complex (see Figure 2.6). This shall be further discussed in the context of locally and laterally reacting materials in Section 2.3.3.

2.3.2 Surface Impedance, Reflection Factor and Absorption Coefficient

A common quantity to describe the acoustical property of a boundary is the surface impedance $\underline{Z}_S$, which is defined as the ratio of sound pressure and normal particle velocity at the surface

$$\underline{Z}_S = \frac{\underline{p}}{\underline{v}_z}\bigg|_{z=0} . \qquad (2.56)$$

The normalized, or *specific*, surface impedance $\underline{\zeta}_S$ is defined as

$$\underline{\zeta}_S = \frac{\underline{Z}_S}{Z_0} . \qquad (2.57)$$

Applying the definition of the surface impedance from Eq. (2.56) to the case of an incoming plane wave in Eq. (2.12), the connection between the (characteristic) surface impedance and the reflection factor for plane waves $\underline{R}$ is obtained:

$$\underline{R}(\theta) = \frac{\underline{\zeta}_S - \frac{1}{\cos(\theta)}}{\underline{\zeta}_S + \frac{1}{\cos(\theta)}} , \qquad (2.58)$$

where the factor $\cos(\theta)$ represents the dependency of the normal particle velocity on the angle of incidence. Eq. (2.58) shows that the reflection factor is always angle-dependent, regardless of the surface impedance itself being angle-dependent (as it is the case in laterally reacting materials) or not.

Another important descriptor for sound reflection at a boundary is the absorption coefficient α. Opposed to the reflection factor, it is a quantity connected to energy rather than amplitude. The absorption coefficient is defined by the part of the sound energy that is not reflected (i.e. absorbed) relative to the incident energy E_{inc}:

$$\alpha(\theta) = \frac{E_{\text{abs}}}{E_{\text{inc}}} = 1 - \frac{E_{\text{refl}}}{E_{\text{inc}}} = 1 - |\underline{R}(\theta)|^2 . \qquad (2.59)$$

Physically meaningful values of the absorption coefficient lie in the range $0 \leq \alpha(\theta) \leq 1$. By splitting the specific surface impedance $\underline{\zeta}_S$ into its real part X_S and imaginary part Y_S and inserting into Eq. (2.58) and Eq. (2.59)

$$\alpha(\theta) = 1 - |\underline{R}(\theta)|^2 = \frac{4 \cdot \frac{1}{\cos(\theta)} \cdot X_S}{\left(X_S + \frac{1}{\cos(\theta)}\right)^2 + Y_S^2}, \tag{2.60}$$

the following important conditions concerning the surface impedance are obtained for $0 \leq \theta \leq \pi/2$:

$$\alpha(\theta) \geq 0 \Rightarrow X_S \in \mathbb{R}_{\geq 0}$$

$$\alpha(\theta) \leq 1 \Rightarrow \left(X_S - \frac{1}{\cos(\theta)}\right)^2 + Y_S^2 \geq 0 \Rightarrow Y_S \in \mathbb{R}$$

$$\Rightarrow -\pi/2 \leq \arg(\underline{Z}_S) \leq \pi/2$$

Here, $\arg(\dots)$ denotes the phase angle of a complex variable.

It follows that the real part of the surface impedance is always positive and hence that the phase of the surface impedance is bounded to the range $[-\pi/2, \pi/2]$. This is a proof that the broadband surface impedance as well as its inverse, the surface admittance

$$\underline{Y}_S = \frac{1}{\underline{Z}_S} = \frac{\underline{Z}_S^*}{|\underline{Z}_S|^2} \tag{2.61}$$

can be considered as minimum-phase — i.e. stable and causal — filters [88, Section 5.6]. Here, $(\dots)^*$ denotes the complex conjugate operator. For the rather theoretical extreme cases of sound hard ($\underline{Z}_S \to \infty$) and soft boundaries ($\underline{Y}_S \to \infty$), with $\alpha = 0$ in both cases, stability is certainly not given. However, in real situations absorption coefficients equal to zero are not encountered. For any $\alpha > 0$, both the impedance as well as the admittance are bounded and hence stability is ensured.

It would thus suffice to obtain either the magnitude or phase of the surface impedance and the other quantity could be obtained by use of the Hilbert transform [88, Section 11.1]. However, this is only possible for the surface impedance; the reflection factor is not minimum-phase and it is especially impossible to obtain the complex surface impedance from a measurement of the absorption coefficient.

2.3.3 Surface Impedance of Locally and Laterally Reacting Materials

So far the surface impedance was considered as a relatively abstract quantity. It will be shown in this section how the surface impedance relates to material parameters and especially how it relates to different classes of absorbers.

The material parameters considered in this section, which are assumed to be known, are the propagation constant $\underline{k}$ and the density ρ, both possibly being complex quantities. It is not within the scope of this thesis how to determine $\underline{k}$ and ρ, either through measurements or using empirically determined models. However, for the purpose of illustration of some basic concepts and for the validation of measurements, the empiric models by DELANY and BAZLEY [89], MIKI [90] and KOMATSU [91] related to fibrous materials will be applied.

In order to determine the surface impedance, the continuity relations of sound pressure and particle velocity at both sides of the boundary are considered, where values in air ($z > 0$) have the index 0 and values in the other medium ($z < 0$) have the index 1:

$$\underline{p}_0\big|_{z=0} \overset{!}{=} \underline{p}_1\big|_{z=0} \quad \wedge \quad \underline{v}_{z,0}\big|_{z=0} \overset{!}{=} \underline{v}_{z,1}\big|_{z=0} \; .$$

If the complex refraction angle in Eq. (2.55) is equal to zero, i.e. the wave in the lower half-space is completely refracted in normal direction, the continuity conditions can be simplified to relate only the local field impedances at both sides of the boundary:

$$\frac{\underline{p}_0}{\underline{v}_{z,0}}\bigg|_{z=0} \overset{!}{=} \frac{\underline{p}_1}{\underline{v}_{z,1}}\bigg|_{z=0} \; .$$

A medium that behaves this way is called *locally reacting* because no propagating wave parallel to the surface exists and hence no pressure exchange between different positions along the boundary is possible. In this case the surface impedance is equivalent to the characteristic impedance of the medium

$$\underline{Z}_S = \underline{Z}_1 \; , \tag{2.62}$$

as long as the medium in the lower half-space is infinitely extended ($z \to -\infty$). A more practical situation consists of an absorber medium backed by a rigid and perfectly reflecting ($\underline{R} = 1$) plane at $z = -d$, in which case the surface impedance can be calculated by[5]

$$\underline{Z}_S = \underline{Z}_1 \coth{(\mathrm{j}\underline{k}_1 d)} \; . \tag{2.63}$$

---

[5]Eq. (2.63) follows from transmission line theory, see for example [70, Section 8.2]

In both cases the surface impedance is independent of the specific wave field above the boundary and it is especially independent of the angle of incidence. Figure 2.7 presents an example of the surface impedance as a function of frequency, calculated for a porous absorber in free air (Eq. (2.62)) and with a rigid backing (Eq. (2.63)). The absorber is characterized by the thickness of $d = 25$ mm and flow resistivity of $\Phi = 5$ kPa s/m^2. The material parameters were calculated with the model by MIKI [90].

It can be confirmed that the phase of the surface impedance is indeed bounded to $[-\pi/2, \pi/2]$ as shown before. For the case with rigid backing (red dashed curve in Figure 2.7), resonant behavior at medium and high frequencies can be observed. At low frequencies, the data shows spring-like behavior ($Z \propto \frac{1}{j\omega}$). This suggests that the absorber is not effective in this frequency range and the acoustical behavior can simply be modeled by a layer of air of thickness d.

This can be confirmed through comparison to the impedance of a layer of air (yellow dot-dashed line in Figure 2.7), calculated by setting $\underline{Z}_1 = Z_0$ and $\underline{k}_1 = k_0$ in Eq. (2.63). At low frequencies the curves are very similar whereas at higher frequencies the damping of the resonances cannot be seen in the impedance of the air layer, which is assumed to be lossless. The slightly shifted resonance frequencies are due to the different speed of sound in the absorber medium.

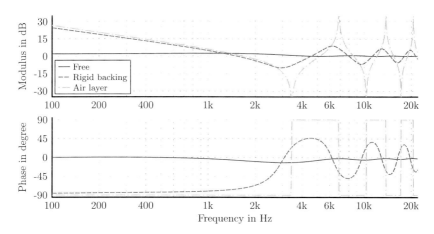

Figure 2.7: Example of the surface impedance of a locally reacting material (thickness $d = 25$ mm, flow resistivity $\Phi = 5$ kPa s/m^2), calculated with the Miki model for porous absorbers without backing (free) and with rigid backing; additionally the surface impedance for a layer of air of the same thickness is shown

The counterpart of a locally reacting material is called *laterally reacting*, where wave propagation parallel to the boundary is possible within the material. The demand for continuity of sound pressure and particle velocity can then be converted to the previously used equality of the field impedances with an additional condition of equal tangential propagation components, which is Snell's law already encountered in Eq. (2.55) in Section 2.3.1. This leads to the equation for the surface impedance of a laterally reacting material of infinite extent

$$\underline{Z}_S(\underline{\theta}') = \frac{\underline{Z}_1}{\cos(\underline{\theta}')} = \frac{\underline{Z}_1}{\sqrt{1 - \left(\sin(\theta) \cdot \frac{k_0}{\underline{k}_1}\right)^2}} \; . \tag{2.64}$$

Eq. (2.64) shows that the surface impedance now also depends on the angle of incidence in air. If the absorbing material is backed by a perfectly reflecting plane at $z = -d$ (compare Eq. (2.63)) the surface impedance becomes

$$\underline{Z}_S(\underline{\theta}') = \frac{\underline{Z}_1}{\cos(\underline{\theta}')} \coth\left(j\underline{k}_1 d \cdot \cos\left(\underline{\theta}'\right)\right) \; , \tag{2.65}$$

where the complex refraction angle $\underline{\theta}'$ can be eliminated as in Eq. (2.64). Figure 2.8 presents a comparison of the surface impedance for a locally and a laterally reacting porous absorber with the same material parameters for different angles of incidence θ. It becomes clear that the behavior of the laterally reacting absorber for large angles of incidence is significantly different from that of a locally reacting one.

In this section, only the lateral behavior of homogeneous porous absorber materials was described as it is the main point of interest in the remainder of this thesis. There are of course other types of laterally reacting surfaces, the most important one being plates excited by bending waves. Their description, however, is not within the scope of this work.

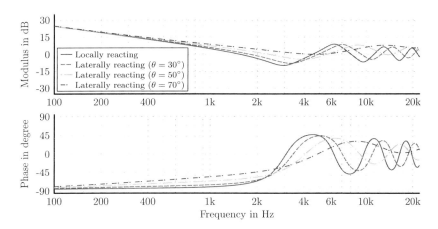

Figure 2.8: Comparison of the surface impedance of a locally and a laterally reacting porous absorber ($d = 25$ mm, $\Phi = 5$ kPa s/m²), calculated with the Miki model for different angles of incidence θ

2.3.4 Reflection of Plane and Spherical Waves

After introducing the concepts of surface impedance and reflection factor the sound field above a boundary with a specific surface impedance can now be described. It is obvious that the sound pressure depends on the type of incident wave (plane or spherical) and on the position of the receiver relative to the origin, here denoted by $\vec{r}_{\mathrm{rec}}$. This dependency is described in the following paragraphs.

Plane Waves   First, the plane wave model will be employed for a simplified description. The geometric setup is depicted in Figure 2.9a, with the wave direction defined by the propagation vector $\vec{k}_{\mathrm{inc}}$. As stated before, the direction of the reflected wave, defined by $\vec{k}_{\mathrm{refl}}$, follows the law of reflection and can hence be found by mirroring the incident propagation vector at the boundary with normal vector $\vec{n}_z$:

$$\vec{k}_{\mathrm{refl}} = \vec{k}_{\mathrm{inc}} - 2\left(\vec{k}_{\mathrm{inc}}\,\vec{n}_z\right)\vec{n}_z\,. \tag{2.66}$$

31

The total sound pressure at the receiver due to the superposition of incident and reflected wave relative to the amplitude $\hat{p}$ of the incident wave is then

$$\frac{\underline{p}_{\text{tot}}\left(\vec{r}_{\text{rec}}, \vec{k}_{\text{inc}}, \vec{k}_{\text{refl}}\right)}{\hat{p}} = e^{-j\vec{k}_{\text{inc}}\vec{r}_{\text{rec}}} + \underline{R}\left(\theta\right)e^{-j\vec{k}_{\text{refl}}\vec{r}_{\text{rec}}}, \qquad (2.67)$$

with the plane wave reflection factor $\underline{R}(\theta)$ as defined in Eq. (2.58).

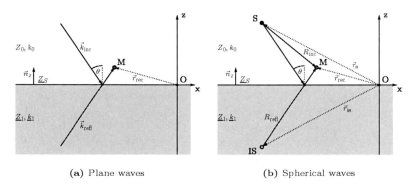

(a) Plane waves   (b) Spherical waves

Figure 2.9: Schematic for sound reflection of plane and spherical waves at the boundary between two media with different acoustic properties

Spherical waves, exact solutions: Complex Image Source Method (CISM)
In the case of spherical waves the situation becomes much more complex as the angle of incidence is no longer constant along the boundary (as is the case for plane waves) and the sound pressure is also determined by the distance between source and receiver (Figure 2.9b shows the geometrical setup with the source and receiver positions). This leads to the situation that the reflected wave is not necessarily a spherical wave, it is additionally composed of surface waves — i.e. waves traveling along the boundary whose amplitudes decay in the normal direction of the surface — as pointed out by LINDELL and ALANEN [92, 93, 94].

The exact solution for the sound pressure field can be obtained by the Laplace transform of an image source distribution $s(q)$ with complex positions of the image sources [95]:

$$\frac{\underline{p}_{\text{tot}}\left(\vec{r}_{\text{s}}, \vec{r}_{\text{rec}}, k\right)}{j\omega\rho_0\hat{Q}} = \underline{G}(k, R_{\text{inc}}) + \int_0^\infty \underline{s}(q) \cdot \underline{G}(k, R_{\text{cis}}(q))\,dq \qquad (2.68)$$

with $R_{\mathrm{inc}} = \|\vec{r}_{\mathrm{rec}} - \vec{r}_s\|$ as the distance between a receiver at $\vec{r}_{\mathrm{rec}}$ and a source at $\vec{r}_s$. $R_{\mathrm{cis}}(q) = \|\vec{r}_{\mathrm{rec}} - \vec{r}_{\mathrm{cis}}(q)\|$ is the distance between the receiver and an image source located at a complex position along the z-axis $\vec{r}_{\mathrm{cis}}(q) = \vec{r}_{\mathrm{is}} + \mathrm{j}\epsilon q \vec{e}_z$. The real part of the location of the image source $\vec{r}_{\mathrm{is}}$ is found by mirroring the source position $\vec{r}_s$ at the boundary normal (compare Eq. (2.66)). $\underline{G}(k,r)$ is the Green's function introduced in Eq. (2.18) and $\vec{e}_z$ is the unit vector in z-direction (as the normal vector to the reflecting surface).

For a locally reacting material $\epsilon = 1$ and

$$\underline{s}(q) = \delta(q) - 2\frac{k}{\underline{\zeta}_S}\mathrm{e}^{-\frac{k}{\underline{\zeta}_S}q}, \tag{2.69}$$

where $\delta(q)$ is the Dirac delta, so that Eq. (2.68) becomes

$$\frac{\underline{p}_{\mathrm{tot}}(\vec{r}_s, \vec{r}_{\mathrm{rec}}, k)}{\mathrm{j}\omega\rho_0\hat{Q}} = \underline{G}(k, R_{\mathrm{inc}}) + \underline{G}(k, R_{\mathrm{refl}}) - 2\frac{k}{\underline{\zeta}_S}\int_0^\infty \mathrm{e}^{-\frac{k}{\underline{\zeta}_S}q} \cdot \underline{G}(k, R_{\mathrm{cis}}(q))\,\mathrm{d}q. \tag{2.70}$$

It can be seen that the pressure field above the surface is composed of contributions by the direct source (at distance R_{inc}), an image source (at distance $R_{\mathrm{refl}} = \|\vec{r}_{\mathrm{rec}} - \vec{r}_{\mathrm{is}}\|$) and an additional integral over a line of image sources at complex locations extending towards infinity. The topic was also investigated by OCHMANN [96] and the approach was extended to yield closed-form solutions in the time-domain [97], which are especially useful for numerical calculations above impedance planes, e. g. for the simulation of outdoor sound propagation.

In their paper, DI and GILBERT [95] give an approximate upper integration limit q_{max} for the integral in Eq. (2.70), which assures that the real exponent in the integrand has decayed approximately 54 dB from the maximum value:

$$q_{\mathrm{max}} = \lambda \cdot \frac{|\underline{\zeta}_S|^2}{\mathrm{Re}\left(\underline{\zeta}_S\right)} = \lambda \cdot \frac{|\underline{\zeta}_S|}{\cos\left(\left|\arg\left(\underline{\zeta}_S\right)\right|\right)}, \tag{2.71}$$

where $\mathrm{Re}\left(\dots\right)$ denotes the real part of a complex variable. The upper integration limit was derived to speed up integral calculations, however most modern numerical integration routines (as e. g. quad/integral in MATLAB) have built-in approximations for limits that tend towards infinity, so that using the pre-computed limit in Eq. (2.71) actually takes longer. Nonetheless, q_{max} is used later on to derive the plane wave approximation.

In analogy to Eq. (2.67) a spherical reflection factor $\underline{Q}(\theta)$ can be defined, which represents the very common mirror-source approach to model sound reflection at surfaces [98]. The total sound pressure relative to the source amplitude term is then written as

$$\frac{\underline{p}_{\text{tot}}\,(\vec{r}_{\text{s}}, \vec{r}_{\text{rec}}, k)}{j\omega\rho_0\hat{Q}} = \underline{G}(k, R_{\text{inc}}) + \underline{Q}\,(k, \theta) \cdot \underline{G}(k, R_{\text{refl}})\,. \qquad (2.72)$$

For the exact solution in Eq. (2.70) the spherical reflection factor becomes

$$\underline{Q}(k, \theta) = 1 - 2\frac{k}{\underline{\varsigma}_S} \int_0^\infty e^{-\frac{k}{\underline{\varsigma}_S}q} \cdot \frac{\underline{G}(k, R_{\text{cis}}(q))}{\underline{G}(k, R_{\text{refl}})}\,\mathrm{d}q\,. \qquad (2.73)$$

Figure 2.10 shows an example of the sound pressure level distribution at 2 kHz due to the reflection of waves from a locally reacting absorber with a surface impedance as in Figure 2.7 (dashed red curve). In Figure 2.10a the case of a plane wave impinging at $\theta = 33°$, with the wave direction indicated by the arrow, is presented and in Figure 2.10b the sound pressure level for the reflection of a spherical wave generated by a point source is graphed, where the position of the point source is marked by a white circle. The setup is based on the schematics shown in Figure 2.9. The sound pressure has been calculated according to Eq. (2.67) and Eq. (2.70), respectively, and the source levels for the plane wave and the point source have been normalized to 1 Pa (at 1 m) for the incident field. The height of the point source above the absorber surface is $z_s = 1.3$ m.

For the plane wave case in Figure 2.10a, a standing wave in normal direction to the reflecting surface is obviously generated by the superposition of incident and reflected wave. In comparison, the distance-dependency of the sound pressure level due to spherical waves can be observed in Figure 2.10b, yielding a more natural model than the rather theoretical construct of plane waves. The plane wave model is thus disregarded in the rest of this thesis.

(a) Plane wave reflection (incident wave direction marked by the black arrow)

(b) Spherical wave reflection (source position marked by a white circle)

Figure 2.10: Example of the sound pressure level distribution for plane and spherical wave reflection at $2\,\mathrm{kHz}$ above a locally reacting absorber plane (black line at the bottom, surface impedance as in Figure 2.7, dashed red curve)

It is now interesting to study the limiting cases of Eq. (2.73) concerning reflection angle and specific surface impedance $\underline{\zeta}_S$, which will establish the relationship to the plane wave reflection factor as a simplified model. For the limit of $\zeta_S \to \infty$ and $\zeta_S = 0$ the spherical reflection factor gives exactly the same result as the plane wave reflection factor according to Eq. (2.58), which is $\underline{Q} = \underline{R} = 1$ and $\underline{Q} = \underline{R} = -1$, respectively. This means that for purely sound hard or sound soft surfaces, the sound pressure is determined only by the original source and an image source, both radiating perfectly spherical waves.

An approximation of Eq. (2.73) for normal incidence is given in [95] as

$$\underline{Q}(0) = \frac{\underline{\zeta}_S - 1}{\underline{\zeta}_S + 1} = \underline{R}(0) \,, \tag{2.74}$$

which, again, is exactly the plane wave reflection factor for normal sound incidence. The condition for this approximation is connected to $q_{\max}$ defined in Eq. (2.71) and the sum of heights of the source and receiver above the boundary (see [95, Section B.III]):

$$z_s + z_{\text{rec}} \gg \frac{q_{\max}}{|\underline{\zeta}_S|}$$

$$\Rightarrow \frac{z_s + z_{\text{rec}}}{\lambda} \cdot \cos\left(\left|\arg\left(\underline{\zeta}_S\right)\right|\right) \gg 1 \,. \tag{2.75}$$

It can be found that this approximation is actually not only valid for normal incidence. This will be evaluated numerically in Chapter 4.

Eq. (2.74) and Eq. (2.75) show that as long as the source and receiver are located several wavelengths away from the reflecting surface and the phase angle of the surface impedance is not too large, then the plane wave reflection factor can be used in the image-source model in Eq. (2.72). This especially makes the simplified model appropriate for the application in geometrical room acoustics [98, 99].

However, for setups with sources and receivers close to the surface, as they usually occur in outdoor propagation scenarios and measurement setups of the reflection factor in the free-field, the spherical reflection factor of the complex image-source model in Eq. (2.73) has to be used. The application and the error due to approximations will be discussed in Chapter 4.

For the sake of completeness, the exact solution of Eq. (2.68) for an infinitely extended laterally reacting material is also presented here. In this case, $\epsilon^{-1} = \sqrt{\underline{k}_1^2 - k_0^2}$ and the source distribution becomes [93]

$$s(q) = \frac{\underline{\rho}_1 - \rho_0}{\underline{\rho}_1 + \rho_0}\delta(q) - \gamma\frac{\underline{\rho}_1}{\rho_0}\int_0^\infty e^{-\gamma q'}\left(\frac{J_2(q'+q)}{q'+q} - \frac{J_2(q'-q)}{q'-q}\right)\,dq',\quad (2.76)$$

with $\gamma = \frac{\rho_0}{\sqrt{\underline{\rho}_1 - \rho_0}}$. Here, ρ_0 and $\underline{\rho}_1$ are the densities in air and in the medium in the lower half-space, respectively. $J_2(x)$ is the Bessel function of the first kind of order two. Eq. (2.76) is especially interesting in the context of numerical simulations, because it could be used to model the reflection from laterally reacting porous absorbers based only on two material parameters. To the knowledge of the author, this has not been done yet.

Spherical waves, exact solutions: Sommerfeld Integral Solution (SIS)
Another commonly used model for the problem of a point source above the boundary with a homogeneous and locally reacting material is the solution originally investigated by Sommerfeld [100] and discussed by various authors. Again, Mechel [87, Chapter 13] gives a very good overview about the different solutions and the implications for numerical evaluation. In any case, using cylindrical coordinates with $r' = \sqrt{(x_{\text{rec}} - x_{\text{s}})^2 + (y_{\text{rec}} - y_{\text{s}})^2}$ together with z_{rec} and z_{s}, the expression for the total sound pressure (compare Eq. (2.70)) becomes[6]

$$\frac{\underline{p}_{\text{tot}}\left(\vec{r}_{\text{s}}, \vec{r}_{\text{rec}}, k\right)}{j\omega\rho_0\hat{Q}} = \underline{G}(k, R_{\text{inc}})$$

$$-j\frac{k}{4\pi}\int_0^{\frac{\pi}{2}+j\infty} J_0\left(kr'\sin\left(\underline{\theta}\right)\right) e^{-jk(z_{\text{s}}+z_{\text{rec}})\cos(\underline{\theta})}\underline{R}\left(\underline{\theta}\right)\sin\left(\underline{\theta}\right)\,d\underline{\theta},$$

$$(2.77)$$

where $J_0(x)$ is the Bessel function of the first kind of order zero. The integration of the second term in Eq. (2.77) is carried out in the complex plane of the angle of incidence $\underline{\theta} = \theta' + j\theta''$, and hence an appropriate path of integration has to be chosen.

---

[6]This is the corrected version of the equation also presented in [3, Section 3.2.3]

Following MECHEL [87], and introducing a change of variables for better numerical evaluation

$$y' = \cos\left(\theta'\right),$$
$$y'' = \sinh\left(\theta''\right),$$
$$\underline{R}\left(y'\right) = \frac{\zeta_S - \frac{1}{y'}}{\zeta_S + \frac{1}{y'}},$$
$$\underline{R}\left(y''\right) = \frac{\zeta_S - \mathrm{j}\frac{1}{y''}}{\zeta_S + \mathrm{j}\frac{1}{y''}},$$

the reflected sound pressure term can be evaluated as

$$\frac{\underline{p}_{\mathrm{refl}}\left(\vec{r}_{\mathrm{s}}, \vec{r}_{\mathrm{rec}}, k\right)}{\mathrm{j}\omega\rho_0 \hat{Q}} = -\mathrm{j}\frac{k}{4\pi} \int_0^1 J_0\left(kr'\sqrt{1-y'^2}\right) e^{-\mathrm{j}k(z_{\mathrm{s}}+z_{\mathrm{rec}})y'}\underline{R}\left(y'\right) \, \mathrm{d}y'$$
$$+ \frac{k}{4\pi}\int_0^\infty J_0\left(kr'\sqrt{1+y''^2}\right) e^{-k(z_{\mathrm{s}}+z_{\mathrm{rec}})y''}\underline{R}\left(y''\right) \, \mathrm{d}y''.$$

$$(2.78)$$

The reflected sound pressure calculated according to Eq. (2.78) yields exactly the same result (within numerical precision) as that of the second and third term in Eq. (2.70). It has been pointed out by different authors that Eq. (2.78) is numerically difficult to evaluate (see the literature review in [87, Section 13.2]). To give an example, using Eq. (2.78) takes roughly 6 times longer compared to Eq. (2.70) for the same setup. Further discussions will be carried out in Chapter 4.

Spherical waves, approximation: Error Function Solution (EFS) A commonly used approximation for locally reacting materials has been described and applied by (among others) INGARD [101], LAWHEAD and RUDNICK [102], and CHIEN and SOROKA [103] and NOCKE et al. [104]. The solution involves the complimentary error function $\mathrm{erfc}(x)$, with which the spherical reflection factor can be calculated by

$$\underline{Q}(\theta) = \underline{R}(\theta) + (1 - \underline{R}(\theta))\left[1 - \mathrm{j}\sqrt{\pi}\underline{w}(\theta)e^{-\underline{w}(\theta)^2}\,\mathrm{erfc}\left(\mathrm{j}\underline{w}(\theta)\right)\right], \quad (2.79)$$

with the numerical distance

$$\underline{w}(\theta) = \frac{1 - \mathrm{j}}{2} \cdot \sqrt{kR_{\mathrm{refl}}} \, \frac{\frac{1}{\zeta_S} + \cos{(\theta)}}{\sqrt{1 + \frac{1}{\zeta_S} \cos{(\theta)}}} \,. \qquad (2.80)$$

The phase angle of the numerical distance has to be restricted to a certain range [105]. With the sign convention used in this work, the definition is written as

$$\frac{\pi}{4} < |\arg{(\mathrm{j}\underline{w}(\theta))}| < \frac{3\pi}{4} \,. \qquad (2.81)$$

This approximation is numerically stable and computationally very fast (it is approximately 2000 times faster than Eq. (2.78) and 400 times faster than Eq. (2.70)). The model has been derived for grazing sound incidence — i.e. for sources and receivers close to the surface — but it also gives acceptable results for other angles of incidence. This will be analyzed in Chapter 4.

2.3.5 Random Sound Incidence

The theoretical formulation of a single reflection at a planar boundary is the basis to establish the description of sound propagation in rooms, where sound is reflected many times at different boundaries with different surface impedances. The multitude of sound reflections leads to the fact that waves will impinge on a specific surface at various angles of incidence, and hence it is of interest to describe the effective absorption for such a sound field.

The commonly used term for this type of sound incidence is *diffuse* (or *random*) and the resulting absorption coefficient is the diffuse incidence absorption coefficient α_{diff}, which can be calculated from the directional absorption coefficient $\alpha(\theta)$ according to PARIS [106]:

$$\alpha_{\mathrm{diff}} = 2 \cdot \int_0^{\frac{\pi}{2}} \alpha(\theta) \cos(\theta) \sin(\theta) \, \mathrm{d}\theta \,, \qquad (2.82)$$

where dependency on the azimuth angle φ is neglected and the factor $\sin(\theta) \, \mathrm{d}\theta$ relates to the solid angle. MAKITA and HIDAKA [107] suggested a revision of the factor $\cos(\theta)$ connected to the angular distribution of the incident sound field. However the experimental validation in [108] has not shown promising results, hence the revised law is not considered here.

The integration in Eq. (2.82) is performed for all theoretically possible angles of incidence, however in practice it is unrealistic to expect sound incidence that is almost parallel to the surface. With reference to the practical application in building acoustics [109, Section 4.3.1.2.1], the maximum angle of incidence can be reduced to $78° \triangleq \frac{\sqrt{3}}{4}\pi$, to yield a more realistic value, which will be called the *field incidence* absorption coefficient following ARETZ and VORLÄNDER [110]:

$$\alpha_{\text{field}} = \frac{\int\limits_0^{\frac{\sqrt{3}}{4}\pi} \alpha(\theta) \sin(2\theta)\, \mathrm{d}\theta}{\int\limits_0^{\frac{\sqrt{3}}{4}\pi} \sin(2\theta)\, \mathrm{d}\theta}, \tag{2.83}$$

where the relation $2\cos(\theta)\sin(\theta) = \sin(2\theta)$ has been used.

Due to the factor $\sin(2\theta)$ in Eq. (2.82) and Eq. (2.83) most weight is given to the value for incident angles around 45 degrees. It has hence been suggested to only measure for this angle — as it is done for sound insulation in ISO 140-5 [111] — when time is limited and minimum effort is desired. In terms of diffuse field absorption this can give a representative result; for an angle-dependent material characterization this is certainly not sufficient if the material is laterally reacting, since the surface impedance becomes angle-dependent and multiple measurements are necessary.

Figure 2.11 shows an example of the directional and the diffuse incidence absorption coefficient corresponding to the surface impedance of the (laterally reacting) porous material described before and plotted in Figure 2.8. Depicted are the curves for normal incidence, 45 degree (*angled*) incidence and for diffuse incidence. For the latter case, the diffuse and field incidence values according to Eq. (2.82) and Eq. (2.83), respectively, are shown for the laterally reacting material. Additionally, the field incidence absorption coefficient for a locally reacting material with the same parameters is presented.

It can be seen that the curve for 45 degree incidence is very close to the one for diffuse incidence and that the field incidence absorption coefficient is slightly higher by about 5 percent. The result for the locally reacting material shows a much more resonant behavior in comparison to the relatively smooth curves for lateral reaction. It can thus be concluded that the behavior of porous absorbers is very different between local and lateral reaction and that the result at 45° incidence is indeed representative of the one for diffuse incidence.

From the discussion on locally reacting materials it becomes evident that for such materials the diffuse or field incidence absorption coefficient can be easily determined if the absorption coefficient (or the surface impedance) for a single angle of incidence is known, since the surface impedance itself does not show any dependency on the angle of incidence. It is in that case actually possible to derive a closed-form solution to Eq. (2.82) and Eq. (2.83), depending only on the real and imaginary part of the surface impedance (see [112, Section 2.5, Eq. (2.42)]).

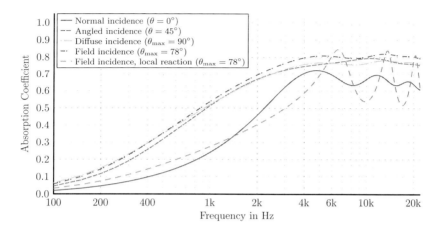

Figure 2.11: Comparison of the absorption coefficient of a laterally reacting porous absorber (with a surface impedance as shown in Figure 2.8) for various types of sound incidence

2.4 Sound Scattering

In the previous sections the specular reflection at surfaces was described and how it relates to material parameters. Sound is, however, not only reflected specularly but energy is also directed away from the specular direction. This process, which depends on the geometrical shape of the surface, is called *scattering*.

The surface roughness (or corrugation) influences the amount of scattering that will occur and this roughness of course has to be seen in comparison to the wavelength. Figure 2.12 schematically shows the differentiation between specular reflection (Figure 2.12a), where the corrugation is small compared to the wavelength, and diffuse reflection (Figure 2.12b), i.e. scattering, where the wavelength is in the same range or smaller than the corrugation so that reflections occur in (almost) all directions.

41

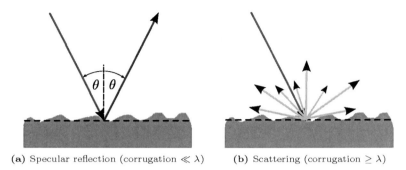

(a) Specular reflection (corrugation $\ll \lambda$)   (b) Scattering (corrugation $\geq \lambda$)

Figure 2.12: Schematic for specular reflection and scattering, depending on the relation between the size of the surface corrugation and the wavelength λ of the incident wave

In this context, surface scattering is generally considered as the redirection of sound energy away from the specular direction. In terms of a quantitative description of this process two coefficients have been suggested and standardized: the scattering coefficient and the diffusion coefficient. Both coefficients use different underlying principles and thus have different applications. The coefficients are obviously frequency-dependent and also angle-dependent. A random-incidence value can be obtained analogous to the absorption coefficient (Eq. (2.3.5)) by the application of Paris' formula in Eq. (2.82).

2.4.1 Scattering Coefficient

The scattering coefficient s is defined as the fraction of energy that is not reflected specularly in relation to the total reflected energy E_{refl} [29]:

$$s = \frac{E_{\text{scat}}}{E_{\text{refl}}} = 1 - \frac{E_{\text{spec}}}{E_{\text{refl}}} \,, \tag{2.84}$$

where $E_{\text{refl}} = (1 - \alpha)E_{\text{inc}}$ is connected to the total incident energy E_{inc} via the absorption coefficient α. The standardized measurement procedure for the random-incidence scattering coefficient in the reverberation chamber is described in ISO 17497-1 [7] (see Section 2.6.2). A method to measure the directional scattering coefficient in the free field has been proposed by MOMMERTZ [113].

From the definition of the (directional) scattering coefficient it becomes clear that large values will be obtained as soon as energy is shifted away from the specular direction, regardless whether the energy is distributed evenly across all directions or whether strong lobing occurs. In terms of the effect for the average sound field in a room this is surely sufficient.

The application of the scattering coefficient lies mostly in geometrical acoustic simulations (i. e. ray-tracing and its derivations), where the angular distribution of the diffusely reflected rays is usually modeled by Lambert's emission law [114], giving an amplitude that depends on the cosine of the angle of sound incidence relative to the surface normal.

2.4.2 Diffusion Coefficient

A different coefficient to describe non-specular reflection at a surface is the diffusion coefficient d, which characterizes the uniformity of the spatially scattered sound. It is thus more intended as a measure of quality of a diffusing surface in distributing the reflected sound energy evenly in all directions.

There are many possible ways to define the diffusion coefficient [67, 115], however — also in terms of standardization — the definition using the spatial cross-correlation has proven most robust and practical. From the reflected sound pressure $\underline{p}_i$ measured at a total of N_p microphones distributed hemispherically above a surface, the directional diffusion coefficient d_θ can be obtained by [116]

$$d_\theta = \frac{\left(\sum_{i=1}^{N_p} |\underline{p}_i|^2 w_i\right)^2 - \sum_{i=1}^{N_p} w_i \left(|\underline{p}_i|^2\right)^2}{\left(\sum_{i=1}^{N_p} w_i - 1\right) \sum_{i=1}^{N_p} w_i \left(|\underline{p}_i|^2\right)^2} , \tag{2.85}$$

where w_i is a weighting coefficient proportional to the surface area sampled by each microphone. Since even a plane surface of finite size yields a non-zero diffusion coefficient due to edge diffraction, the *normalized* diffusion coefficient $d_{\theta,n}$ has been proposed:

$$d_{\theta,n} = \frac{d_\theta - d_{\theta,r}}{1 - d_{\theta,r}} , \tag{2.86}$$

with $d_{\theta,r}$ as the diffusion coefficient for the flat reference surface of the same dimensions as the test sample.

Measurements of the diffusion coefficient are relatively complex and a method in the reverberation chamber still does not exist. Hence, the directional diffusion coefficient has to be found for different angles of incidence. The suggested measurement setup involves a so-called *goniometer*, where source and receiver are moved along arcs with different radii and the test sample is placed in the center of the setup on a turntable. With the focus on application in room acoustics, the diffusion coefficient is usually determined far away from the test sample: AES 4id 2001 [116] suggests a source distance of 10 m and a receiver distance of 5 m. As this is not always feasible, small-scale experiments are often performed. This permits to significantly reduce the size of the measurement setup but it also means that the frequency range has to be extended towards much higher frequencies.

The application of the diffusion coefficient is restricted to the characterization of diffusing surfaces; in contrast to the scattering coefficient it cannot easily be used for sound field simulations as its definition does not provide enough information to model diffuse reflections. On the other hand, the raw data obtained during measurements of the directional diffusion coefficient could be used for a detailed description of sound reflection from a surface, depending on the angle of incidence of the incoming wave. Whether the increased effort required for implementation and data storage in simulation tools is justified in terms of increased simulation accuracy still has to be investigated.

It can thus be concluded that for a complete description of diffuse reflections from surfaces both the scattering coefficient as well as the diffusion coefficient are important and valuable. However from an application point-of-view the scattering coefficient is more suitable to be implemented and it is much easier to obtain measurement data.

2.5 Room Acoustics

After describing the sound reflection at single boundaries the transition can be made from this rather theoretical construct to more applicable situations as they are found in rooms, or more generally in enclosed spaces. In this section the foundation for statistical room acoustics will be briefly laid out to obtain the equations needed for measurement methods in the reverberation chamber.

Solving the inhomogeneous Helmholtz equation with appropriate boundary conditions, related to the reflective properties of the room walls, leads to a general equation for the sound pressure transfer function $\underline{H}_p(f)$ from a monopole source to a monopole receiver in a room [112, Section 3.1]

$$\underline{H}_p(f) = \sum_n = \frac{P_n(f)}{f^2 - f_n^2 - \mathrm{j}\,\frac{\delta_n}{\pi}\,f_n}\,, \tag{2.87}$$

where $\underline{P}_n(f)$ is a factor related to the position and properties of the source and the receiver and to the room boundaries. f_n and δ_n are the eigenfrequency and damping constant, respectively, related to a room mode and it is assumed that $\delta_n \ll 2\pi f_n$, a condition that is generally fulfilled. The damping constant δ_n is related to the resonance half-power bandwidth Δf_n corresponding to f_n by

$$\Delta f_n = \frac{\delta_n}{\pi} . \tag{2.88}$$

The expression for the sound pressure in Eq. (2.87) is basically a superposition of many damped resonances. It can be shown that the number of resonance frequencies (or *eigenfrequencies*) increases with the third power of frequency

$$N_f = \frac{4\pi}{3} V_{\text{room}} \left(\frac{f}{c}\right)^3 , \tag{2.89}$$

and the eigenfrequency density with the second power of frequency

$$\frac{\partial N_f}{\partial f} = 4\pi V_{\text{room}} \frac{f^2}{c^3} , \tag{2.90}$$

where V_{room} is the room volume in m^2 and c is the speed of sound in m/s. These equations are relatively good approximations for most rooms, especially at higher frequencies (see [112, Section 3.2]).

SCHROEDER [117] showed that an approximate transition frequency can be given between the range where the transfer function in a room is determined by a few, well separated decaying modes, and higher frequencies, where many of such modes overlap.[7] By demanding an overlap of three modes within one half-power bandwidth and using Eq. (2.88) and Eq. (2.90), the so-called *Schroeder frequency* f_s is obtained [118]:

$$f_s = \sqrt{\frac{c^3}{4 \ln(10)}} \cdot \sqrt{\frac{T}{V_{\text{room}}}} , \tag{2.91}$$

where T is the reverberation time in seconds, one of the fundamental quantities in room acoustics. The reverberation time T is defined as the time it takes for the steady-state sound energy $E(t)$ in a room to decrease to a millionth of its initial value after the sound source has been switched off.

---

[7]In the original publication, SCHROEDER [117] considered an overlap of 10 normal modes within one half-power bandwidth in the same frequency range; this was later compared to experimental results and the Schroeder frequency as it is used today was defined for 3 overlapping modes.

Figure 2.13 shows an example of the transfer function calculated according to Eq. (2.87) for a room of $V_{\mathrm{room}} = 224\,\mathrm{m}^3$ with a frequency-independent damping constant $\delta = 1.73\,\mathrm{s}^{-1}$. The Schroeder frequency in this case is $f_s = 280\,\mathrm{Hz}$ and it is additionally shown in the graph.

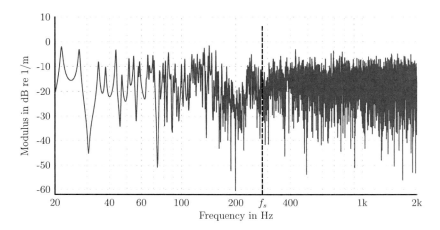

Figure 2.13: Example of an analytic transfer function calculated according to Eq. (2.87) for a room of $V_{\mathrm{room}} = 224\,\mathrm{m}^3$ with a frequency-independent damping constant $\delta = 1.73\,\mathrm{s}^{-1}$; additionally the Schroeder frequency $f_s = 280\,\mathrm{Hz}$ is depicted

It becomes clear that for frequencies below the Schroeder frequency only a few separated resonances can be seen in the transfer function, whereas with increasing frequency their spacing becomes smaller. Above approximately 200 Hz, the level of the transfer function seems to vary randomly, which is due to the high number of overlapping modes with quasi-random phase relations. It should be noted at this point that the Schroeder frequency is only a rough estimate for the transition between modal and statistical sound fields, which can be confirmed in the example where the high modal density is observed for frequencies well below the value of $f_s = 280\,\mathrm{Hz}$. Nonetheless, it serves as a good rule-of-thumb.

With the high modal overlap above the Schroeder frequency, it does not make sense to describe the transfer function (and thus the sound field) in the room based on the individual damping constants δ_n but instead an average value

$< \delta_n >$ within a certain frequency range — typically third-octave bands — can be used, which is related to the reverberation time by

$$< \delta_n >= \frac{3 \ln(10)}{T} .$$ (2.92)

The use of an average value is based on the reasonable assumption that in such a narrow frequency range the damping constants are all more or less equal as the boundary conditions are usually smooth functions with respect to frequency. Under these conditions, the sound energy decay as a function of time t can be expressed by (for the derivation, see [112, Section 5.3])

$$E(t) = E(0) \cdot e^{[\overline{n} \ln(1-\overline{\alpha}) - m \, c] \, t} \qquad t \geq 0 ,$$ (2.93)

with the average number of wall reflections per unit time

$$\overline{n} = \frac{c}{\overline{l}} = \frac{c}{4} \frac{S_{\text{room}}}{V_{\text{room}}} .$$ (2.94)

Here $\overline{l}$ is the so-called *mean free path* [119, 120, 121] and S_{room} is the room surface area in m^2. m is the air attenuation coefficient (see Section 2.1.4) to account for energy loss during the propagation in air. This coefficient depends on the temperature and relative humidity and is especially important for high frequencies and large rooms. $\overline{\alpha}$ is the average absorption coefficient of the room defined by

$$\overline{\alpha} = \frac{1}{S_{\text{room}}} \sum_i S_i \alpha_i ,$$ (2.95)

where the S_i are the individual surface areas in m^2 and α_i the corresponding absorption coefficients.

Using Eq. (2.93) and the definition of the reverberation time a very important equation is obtained, which is known as Eyring's reverberation formula [122]

$$T = \frac{24 \ln(10)}{c} \frac{V_{\text{room}}}{4m V_{\text{room}} - S_{\text{room}} \ln(1 - \overline{\alpha})} ,$$ (2.96)

which connects the reverberation time to the average absorption coefficient in a room. This important relation is used in many practical applications in acoustics. If the average absorption coefficient is sufficiently small, the natural logarithm can be expressed by a Taylor series expansion using only the first-order term [123]:

$$- \ln(1 - x) \approx x, \qquad x \ll 1 ,$$ (2.97)

which gives for the reverberation time

$$T = \frac{24 \ln (10)}{c} \frac{V_{\text{room}}}{A + 4mV_{\text{room}}} , \qquad (2.98)$$

where the *equivalent absorption area* $A = S_{\text{room}} \overline{\alpha}$ has been introduced. This equation is the same as the one derived by SABINE [4], although it was obtained with different reasoning.

In practical situations, usually the reverberation time is measured in order to obtain information about the average absorption in the room, hence it is useful to solve Eq. (2.96) and Eq. (2.98) for $\overline{\alpha}$, giving

$$\overline{\alpha}^{(\text{Eyr})} = 1 - e^{-\frac{4\,V_{\text{room}}}{S_{\text{room}}} \left(\frac{6\ln(10)}{c \cdot T} - m \right)} , \qquad (2.99)$$

for an absorption coefficient according to Eyring's formula and

$$\overline{\alpha}^{(\text{Sab})} = \frac{4\,V_{\text{room}}}{S_{\text{room}}} \left(\frac{6\ln(10)}{c \cdot T} - m \right) , \qquad (2.100)$$

for the Sabine absorption coefficient. In most practical situations including standardized measurement methods (see Section 2.6) the latter equation is used. Obviously both absorption coefficients are related through

$$\overline{\alpha}^{(\text{Eyr})} = 1 - e^{-\overline{\alpha}^{(\text{Sab})}} , \qquad (2.101)$$

so that the values can always be converted. The systematic error introduced by using Sabine's instead of Eyring's formula will be investigated in Chapter 3.

2.6 Standardized Measurement Methods

This section will summarize two very common measurement methods in the reverberation chamber to determine the random-incidence absorption and scattering coefficient. These methods will be analyzed thoroughly with respect to their sensitivity to measurement uncertainties in Chapter 3. In the following, it will be assumed that measurements, especially of reverberation times, are performed with correlation methods using MLS [124] or swept-sine signals [125] instead of the outdated and less reliable interrupted noise method. ISO 18233 [126] can be considered as a reference on the use of modern measurement techniques in standardized methods such as ISO 354 [5], ISO 10140-2 [127] and ISO 17497-1 [7]. The reverberation time is then obtained with the method of the integrated impulse response suggested by SCHROEDER [128].

2.6.1 Random-Incidence Absorption Coefficient (ISO 354)

In order to determine the absorption coefficient for random-incidence α_{sample} of a sample of surface area S_{sample} the reverberation time has to be measured in a sufficiently diffuse sound field, once without and once with the specimen inside.

Using Eq. (2.100) and Eq. (2.95), the absorption coefficient is then given by [5, 6]

$$\alpha_s = \frac{S_{\text{room}}}{S_{\text{sample}}}(\overline{\alpha}_2 - \overline{\alpha}_1),$$
$$= \frac{4\,V_{\text{room}}}{S_{\text{sample}}}\left(\frac{6\ln(10)}{c_2 \cdot T_2} - \frac{6\ln(10)}{c_1 \cdot T_1} - (m_2 - m_1)\right), \qquad (2.102)$$

where all quantities with index 1 have been measured in the empty chamber and the quantities with index 2 have been measured with the test sample inside the chamber. Here, c_i and m_i are determined by measurements of temperature and relative humidity and calculations according to ISO 9613-1 [74].

A situation that occurs once the sample is put inside the measurement chamber is that a part of the room surface area is covered by the sample. In the calculation of the sample absorption coefficient this should be taken into account. ISO 354 does not consider this effect, however in ASTM C423 it is included to give a slightly different equation for α_s:

$$\alpha_{s,\text{ASTM}} = \frac{S_{\text{room}}}{S_{\text{sample}}}(\overline{\alpha}_2 - \overline{\alpha}_1) + \overline{\alpha}_1 = \alpha_{s,\text{ISO}} + \overline{\alpha}_1. \qquad (2.103)$$

The influence of neglecting this correction is usually deemed negligible but it will be investigated in more detail in Chapter 3.

The surface area of the sample is restricted by ISO 354 to be at least $10\,\text{m}^2$ and for rooms up to $V_{\text{room}} = 200\,\text{m}^3$ the maximum allowed value is $S_{\text{sample}} = 12\,\text{m}^2$. For rooms with a volume of more than $200\,\text{m}^3$, the upper limit of the sample surface area has to be multiplied by a factor of $(V_{\text{room}}/200\,\text{m}^3)^{2/3}$. The advice is given that a larger sample should be used for large rooms and for samples with a low absorption coefficient.

The diffuseness of the sound field is usually improved by inserting scattering elements (or *diffusers*) into the chamber, either as hanging panels or as boundary diffusers on the walls. The necessary number of such elements to achieve adequate diffuseness can be checked with procedures given in ISO 354, ASTM C423 and ASTM E90, each with a different approach. The effectiveness of these methods is questionable and has been investigated by

BRADLEY et al. [129]. Nonetheless, adding any scattering elements into the chamber does increase the state of diffuseness.

The absorption of the empty chamber (including the scattering elements) — best described by the equivalent absorption area A — also affects the sound field such that more absorption leads to a less diffuse sound field as energy is constantly drawn towards the absorbing surfaces, violating the demand for isotropy of the sound field. To ensure an adequate diffuseness, ISO 354 gives upper limits for the equivalent absorption area of the empty chamber (see Table 2.1), which are calculated for a reference room volume of $200\,\mathrm{m}^3$ and have to be multiplied by a factor of $(V_{\mathrm{room}}/200\,\mathrm{m}^3)^{2/3}$ if the actual chamber volume deviates from the reference value.

Table 2.1: Limits for the equivalent absorption area of the empty chamber according to ISO 354

Frequency (in Hz)	100–630	800	1000	1250	1600
A_{max} (in m^2)	6.5	6.5	7.0	7.5	8

Frequency (in Hz)	2000	2500	3150	4000	5000
A_{max} (in m^2)	9.5	10.5	12	13	14

2.6.2 Random-Incidence Scattering Coefficient (ISO 17497-1)

The measurement of the scattering coefficient in the reverberation chamber is closely connected to the measurement of absorption coefficients, with two additional measurements required [29]. Figure 2.14 shows a schematic of the measurement procedure as it is described in ISO 17497-1 [7].

Additionally to the measurements needed for the absorption coefficient (Figure 2.14a and Figure 2.14b), the reverberation times also have to be measured for different orientations of the turning table, once without the sample (Figure 2.14c) and with the sample on it (Figure 2.14d). The impulse responses for the last two situations are obtained by phase-locked averaging of 60–80 measurements, either during a continuous or step-wise movement of the turning table.

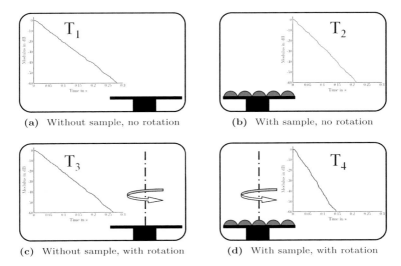

Figure 2.14: Schematic representation of the measurement procedure for scattering coefficients according to ISO 17497-1

By averaging across different orientations of the scattering sample the surface corrugations will be effectively smoothed out so that the determination of the reverberation time T_4 (and in consequence the average absorption coefficient $\overline{\alpha}_4$) from the averaged impulse responses will yield results for the non-scattering part of the sample surface. From the four room absorption coefficients $\overline{\alpha}_1$–$\overline{\alpha}_4$ obtained under these conditions, the scattering coefficient s is calculated by

$$s = \frac{\alpha_{\mathrm{spec}} - \alpha_s}{1 - \alpha_s} = 1 - \frac{1 - \alpha_{\mathrm{spec}}}{1 - \alpha_s}, \qquad (2.104)$$

with

$$\alpha_{\mathrm{spec}} = \frac{S_{\mathrm{room}}}{S_{\mathrm{sample}}} (\overline{\alpha}_4 - \overline{\alpha}_3), \qquad (2.105)$$

as the so-called *specular absorption coefficient*. Eq. (2.104) directly follows from Eq. (2.84).

The measurement of T_3 (Figure 2.14c) would in theory not be necessary as turning the flat and empty baseplate of the turntable would not yield different reverberation times. However, in actual setups slightly tilted or uneven baseplates have a systematic effect also on T_4. In order to quantify this influence the scattering coefficient of the baseplate can be calculated by

$$s_{\text{base}} = \frac{S_{\text{room}}}{S_{\text{sample}}}(\overline{\alpha}_3 - \overline{\alpha}_1) \,, \qquad (2.106)$$

which can be used as a measure of the precision of the setup. ISO 17497-1 gives upper limits of the permissible baseplate scattering in order to assure reliable results (see Table 2.2).

Table 2.2: Limits for baseplate scattering according to ISO 17497-1

Frequency (in Hz)	100–400	500	630	800	1000	1250
$s_{\text{base},max}$	0.05	0.05	0.10	0.10	0.10	0.15

Frequency (in Hz)	1600	2000	2500	3150	4000	5000
$s_{\text{base},max}$	0.15	0.15	0.20	0.20	0.20	0.25

In comparison to ISO 354, ISO 17497-1 provides a simplified upper limit of the equivalent absorption area of the empty chamber A_1 (compare Table 2.1) given by

$$A_1 \leq 0.3 \cdot V^{2/3} \,, \qquad (2.107)$$

which basically means that $\overline{\alpha}_1 \leq 0.045$ for rooms that are more or less rectangularly shaped.

It can be seen from Eq. (2.104) that the sample absorption coefficient α_s also has an influence on the measurement of the scattering coefficient, which is then accounted for. Nonetheless, it is probably evident that if the sample absorbs a lot of the energy, less energy is left for the determination of the scattering properties of the sample. ISO 17497-1 gives an upper limit of $\alpha_s = 0.5$ in order to limit the uncertainty due to the sample absorption. The influence of this factor will be investigated in Chapter 3.

Opposed to ISO 354, where the dynamic range of the decay curve to evaluate the reverberation time is demanded to be at least 20 dB, the dynamic range in ISO 17497-1 is explicitly set to 15 dB, i.e. to obtain T_{15} the decay curve is evaluated between -5 dB and -20 dB. It will become clear during the uncertainty analysis in Section 3.3 that this actually increases the measurement uncertainty due to spatial fluctuations of the reverberation times.

2.7 Uncertainty Analysis

To analyze the standardized measurement methods regarding their susceptibility to uncertainties that may appear in the input quantities, the concept
of uncertainty analysis is very useful. The general idea of investigating (and
reporting) uncertainties that are inherent in any type of measurement has
been investigated and applied for several decades [38] and the concept has
been accepted in all major fields of engineering [39]. The main reference and
de-facto standard in this field is the "Guide to the expression of uncertainty
in measurement (GUM)" [44, 130] and the analysis carried out in this thesis
will in large parts be based on this document.

In connection to uncertainty analysis, some terminology related to the results
of measurements has to be introduced:

- *Accuracy* defines the degree of "trueness", i. e. how close the result of
  a measurement is to the true value. It is hence not always possible to
  actually quantify the accuracy of a test method as the true result is
  hardly ever known.

- *Precision* is a measure of uncertainty of a measurement result, i. e. how
  much variation has to be expected for repeated measurements under
  the same or similar conditions. The precision of measurement methods
  is often determined by intra- or inter-laboratory tests.

- *Repeatability* denotes the precision of measurements performed by the
  same investigator with the same equipment and in the same laboratory.
  This is hence an outcome of intra-laboratory tests and it is mainly
  influenced by the possibilities of the setup of sources, receivers and the
  test sample.

- *Reproducibility* denotes the precision of measurement results obtained by
  different investigators with different equipment at different laboratories.
  It is the outcome of inter-laboratory tests and is influenced by the
  susceptibility of the method to changes in the equipment and the
  measurement environment.

The goal of the first part of this thesis is to analyze the precision of
standardized measurement methods, but it is neither related to repeatability
nor reproducibility. The intention is merely to analyze the influence of the
measurement chamber and of the necessary calculations of the measurement
method on the final result. By doing this, the precision of a single measurement
can be determined analytically. The determination of the precision of a single
measurement is considered the most fundamental part of an analysis of test

methods, which will always affect subsequent intra- or inter-laboratory tests, as they have been often carried out.

It will be assumed in the following that there exists a continuous mathematical relationship between the output quantity y (also called the *measurand*) and the input quantities x_i, i. e.

$$y = f(x_1, x_2, x_3, \ldots, x_N) \ . \tag{2.108}$$

The principle of propagation of uncertainty is then based on the Taylor series expansion of this relationship

$$\Delta f(x) = f(x + \Delta x) - f(x) = \sum_{n=1}^{\infty} \frac{1}{n!} \cdot \frac{\partial^n f(x)}{\partial x^n} \cdot (\Delta x)^n \ , \tag{2.109}$$

where $n!$ is the factorial of n and $\frac{\partial^n f(x)}{\partial x^n}$ is the n-th derivative of f at x. Usually the expansion is stopped after the linear $(n = 1)$ term. Although this greatly facilitates the application, care has to be taken to ensure that the higher-order terms do not contribute significantly to the result.

2.7.1 Systematic Deviations

One possible uncertainty that may affect the outcome of measurements is a systematic deviation of the input quantities, either because of the way they are calculated or because their values are measured incorrectly. With respect to the field of room acoustics an example for the latter type of deviation would be an erroneous value for the room surface area and/or volume as this would affect all calculations based on these parameters equally. A cause for a systematic deviation due to the way of calculating the input quantities could for example be the difference between the Eyring and Sabine absorption formulas (Eq. (2.99) and Eq. (2.100), respectively). The influence of such deviations will be investigated in Section 3.1.2.

It is characteristic of systematic deviations that they are usually known both in sign and magnitude, and hence the effect on the output quantity can be predicted and partly corrected for. From the terminology introduced in the beginning of this section it becomes clear that systematic deviations affect the accuracy of a test method, although this does not necessarily mean that the true result is known. The deviations could also be considered with respect to the theoretically correct result, believed to represent the true value.

By directly applying the linear expansion in Eq. (2.109) to Eq. (2.108) the equation for uncertainty (or error) propagation of systematic deviations is obtained:

$$\Delta y = \sum_{i=1}^{N} \frac{\partial y}{\partial x_i} \cdot \Delta x_i \, , \qquad (2.110)$$

where Δx_i and Δy are the deviations of the input and output quantities, respectively, from the true (or expected) value. $\frac{\partial y}{\partial x_i}$, sometimes also denoted by c_i as the *sensitivity coefficient*, is the partial derivative of y with respect to x_i evaluated for the value of the input quantity considered to be true. The sensitivity coefficient c_i can be regarded as a measure of the susceptibility of the output quantity to the variation of a specific input quantity and it is a key element of uncertainty analysis.

Eq. (2.110) shows that contributions from different input quantities can cancel out and thus lead to smaller deviations in the output if either the partial derivatives or the deviations differ in sign.

2.7.2 Random Deviations

The systematic deviations mentioned in the last subsection only play a minor role concerning measurement uncertainties mostly because they are usually known and can to some extent be corrected for. The major part of the deviations that are encountered during measurements originates from random variations in the input quantities. This is obvious, as there can never be technical measurement equipment that works without the least bit of error. The measurement environment can also introduce uncertainties into the result. Whether these variations are considered large or small depends of course on the context and the quantity under investigation.

Opposed to the case discussed in Section 2.7.1, random deviations of the input quantities do not affect the accuracy of a test method, as such variations could not lead to a higher or lower degree of trueness of the result. Instead, they define the precision of the test method, indicating how reliable the result of one measurement is with regard to repetitions of the same procedure under the same conditions.

Concerning random deviations of the input and output quantities, the analysis is based on the result of a number of M repeated measurements of the same sample involving the mean (or average) value

$$\mu_x = \frac{1}{M} \sum_{m=1}^{M} x_m \, , \qquad (2.111)$$

as an estimate of the true value and the standard deviation

$$\sigma_x = \sqrt{\frac{1}{M-1}\sum_{m=1}^{M}(x_m - \mu_x)^2}\,, \qquad (2.112)$$

as a measure of the variability of the results. The standard deviation is related to the variance by $\mathrm{var}_x = \sigma_x^2$. Finally, the uncertainty u_x, which is the standard deviation of the mean (also called *standard error*), is defined by

$$u_x = \frac{\sigma_x}{\sqrt{M}}\,, \qquad (2.113)$$

and it is a measure of precision of the mean value μ_x.

The uncertainty of the output related to each input quantity follows the law of error propagation [43]

$$u_{y_i} = \frac{\partial y}{\partial x_i} u_{x_i}\,, \qquad (2.114)$$

where the partial derivative is evaluated at the mean value of x_i. The combined uncertainty of the output quantity due to random deviations of N input quantities is then obtained by (compare Eq. (2.110))

$$u_y = \sqrt{\sum_{i=1}^{N} u_{y_i}^2}\,, \qquad (2.115)$$

where the input quantities have been assumed to be uncorrelated. In comparison to Eq. (2.110) this leads to the situation that the contributions from different input quantities never cancel out and hence a kind of worst-case value is calculated for the combined uncertainty. If the correlation coefficient $r(x_i, x_j)$ between the input quantities x_i and x_j is not zero, a second term has to be added to Eq. (2.115) (see [44, Section 5.2]), to obtain

$$u_y = \sqrt{\sum_{i=1}^{N}\left(\frac{\partial y}{\partial x_i}\cdot u_{x_i}\right)^2 + 2\sum_{i=1}^{N-1}\sum_{j=i+1}^{N}\frac{\partial y}{\partial x_i}\frac{\partial y}{\partial x_j}u_{x_i}u_{x_j}r(x_i,x_j)}\,. \qquad (2.116)$$

The correlation coefficient is a symmetric quantity so that $r(x_i, x_j) = r(x_j, x_i)$. Eq. (2.116) shows that the combined uncertainty for correlated quantities can be lower than for completely independent quantities if either one of the partial derivatives or the correlation coefficients are negative.

The uncertainty u_y can be used to provide a confidence interval (with associated confidence level) for expected mean values of the output quantity in repeated measurements as $\mu_y \pm C \cdot u_y$, where C is the so-called *coverage factor*. The coverage factor is related to the assumption of the probability distribution of the output quantity. For the usual (and realistic) case of a normal distribution, $C = 1$ and $C = 1.96$ give confidence intervals at the 67 % and 95 % confidence level, respectively. For the uncertainty analysis in Chapter 3 a confidence level of 95 % and hence a value of $C = 1.96$ will be used.

3

Uncertainty Analysis of Reverberation Chamber Measurements

This chapter will provide a thorough analytical uncertainty analysis of the two most important standardized measurement methods concerning room acoustical applications: ISO 354 related to the random-incidence absorption coefficient and ISO 17497-1 for the random-incidence scattering coefficient. Both of these methods are defined for measurements in the reverberation chamber. The theoretical background that applies here is thus related to statistical room acoustics, as described in Section 2.5.

The basis for both of the standardized methods analyzed in the scope of this work is the measurement of reverberation times in a diffuse sound field and the subsequent calculation of the average room absorption coefficient $\overline{\alpha}$ (see Eq. (2.100)). First, the latter quantity will be investigated concerning its susceptibility to uncertainties and later the results will be applied to the standardized random-incidence coefficients using further propagation of uncertainty.

As already mentioned in Section 2.7, both systematic as well as random deviations are of interest in uncertainty analysis, and hence both will be treated here. Part of the results have already been published in [131, 132, 133, 134], but the analysis presented in this thesis expands these results. The input quantities in the calculation of the absorption and scattering coefficient can be categorized with regard to the type of uncertainty they are usually subject to and the cause for the uncertainty. This categorization is summarized in Table 3.1 and it will be followed for the uncertainty analysis in the rest of this chapter.

From the high number of influencing factors it becomes clear that the following analysis and results are merely a starting point for further studies related to measurement uncertainty. The most important factors that determine the precision of a measurement result in the reverberation chamber will be investigated briefly. However, an extended study in chambers with different dimensions and with a large number of different test samples remains to be done to validate and extend the results obtained in this thesis.

Table 3.1: Uncertainty factors regarding standardized measurements in a reverberation chamber

Uncertain Quantity	Affected Quantity	Type of Uncertainty	(Main) Cause of Uncertainty
V_{room} in m^3	$\overline{\alpha}$	Systematic	Inaccurate Estimation
S_{room} in m^2	$\overline{\alpha}$	Systematic	Inaccurate Estimation
c in m/s	$\overline{\alpha}$	Random	Measurement Uncertainty
m in 1/m	$\overline{\alpha}$	Random	Measurement Uncertainty
T in s	$\overline{\alpha}$	Random	Spatial Variation[1]
$\overline{\alpha}^{(\text{Sab})}$ vs. $\overline{\alpha}^{(\text{Eyr})}$	$\overline{\alpha}$	Systematic	Method of Calculation
S_{sample} in m^2	α_s, s	Systematic	Inaccurate Estimation
S_{room} covered by S_{sample}	α_s, s	Systematic	Neglect of Correction Term

[1]Other potential causes like non-linear decay curves and influence of the measurement equipment are considered negligible here and will not be investigated.

3.1 Average Room Absorption Coefficient

If the reverberation time T has been determined in a room with a diffuse sound field, the average room absorption coefficient of that room can be calculated by either Eq. (2.99) or Eq. (2.100). As the standardized methods always rely on the latter, i. e. the Sabine equation, this will also be considered as the reference.

The first step of an uncertainty analysis is the calculation of the sensitivity coefficients, i. e. the partial derivatives of the function for the output quantity with respect to each input quantity. In the case of the room absorption coefficient as output quantity, the input quantities are

1. the room volume V_{room} and surface area S_{room},

2. the speed of sound c,

3. the air attenuation coefficient m, and

4. the reverberation time T.

The sensitivity coefficients for each of these input quantities will now be calculated and briefly discussed. Examples and verification measurements related to the actual uncertainties will mainly be presented for the standardized coefficients for the absorption coefficient in Section 3.2 and for the scattering coefficient in Section 3.3.

3.1.1 Sensitivity Coefficients

A Room volume and Surface Area

The sensitivity to erroneous values of the room dimensions, calculated by

$$\frac{\partial \overline{\alpha}}{\partial V_{\text{room}}} = \frac{4}{S_{\text{room}}} \cdot \left(\frac{6 \ln(10)}{c \cdot T} - m \right) = \frac{\overline{\alpha}}{V_{\text{room}}}, \tag{3.1}$$

$$\frac{\partial \overline{\alpha}}{\partial S_{\text{room}}} = -\frac{4 V_{\text{room}}}{S_{\text{room}}^2} \cdot \left(\frac{6 \ln(10)}{c \cdot T} - m \right) = -\frac{\overline{\alpha}}{S_{\text{room}}}, \tag{3.2}$$

is always less for larger rooms and for more reverberant rooms. The different sign of the sensitivity coefficients indicates that for an overestimation of the room volume, the room absorption coefficient is also overestimated, whereas for an overestimation of the surface area it is underestimated.

B Speed of Sound

The influence of variations of the speed of sound

$$\frac{\partial \overline{\alpha}}{\partial c} = -\frac{4\,V_{\text{room}}}{S_{\text{room}}} \cdot \frac{6\ln(10)}{c^2 \cdot T} = -\frac{1}{c}\,(\overline{\alpha} + \alpha_{\text{air}}) \tag{3.3}$$

with the apparent absorption caused by air attenuation

$$\alpha_{\text{air}} = \frac{4\,m\,V_{\text{room}}}{S_{\text{room}}} \tag{3.4}$$

is apparently larger for more absorptive rooms (with shorter reverberation times) and it increases with the mean free path, which in turn increases with the room volume. It was shown in Section 2.1.4 how the speed of sound depends on temperature and also that the influence of relative humidity is negligible. This means that the sensitivity coefficient related to the speed of sound can actually be connected to the temperature $\Delta\Theta$ in degree Celsius by applying the chain rule (see Eq. (2.22)):

$$\left.\frac{\partial \overline{\alpha}}{\partial \Delta\Theta}\right|_c = \frac{\partial \overline{\alpha}}{\partial c} \cdot \frac{\partial c}{\partial \Delta\Theta} = -\frac{0.6\,\frac{\text{m}}{\text{s} \cdot {}^\circ\text{C}}}{c} \cdot (\overline{\alpha} + \alpha_{\text{air}})$$

$$= -\frac{1}{552.33\,{}^\circ\text{C} + \Delta\Theta} \cdot (\overline{\alpha} + \alpha_{\text{air}}) \,. \tag{3.5}$$

Eq. (3.5) shows that the influence of variations in temperature on the room absorption coefficient with respect to the speed of sound is usually lower than $\frac{1}{500\,{}^\circ\text{C}}$ in the typical range of temperatures and hence negligible. It has to be noted that temperature variations not only affect the speed of sound but also the air attenuation coefficient and hence this has to be investigated as well.

C Air Attenuation Coefficient

The sensitivity with respect to the air attenuation coefficient

$$\frac{\partial \overline{\alpha}}{\partial m} = -\frac{4\,V_{\text{room}}}{S_{\text{room}}} \tag{3.6}$$

is a constant value only depending on the dimensions of the chamber. Among the sensitivity coefficients presented here, the one related to the air attenuation coefficient is by far the highest. However one has to keep in mind that the coefficient m (and its variation) is orders of magnitude lower than for the other parameters. It will thus be necessary to perform a comparison including typical values of the variation of the input quantities in Section 3.1.3.A. Opposed to the case of the speed of sound, it is not easily possible to analytically determine the influence of temperature and humidity variations on the air attenuation coefficient as the equations in ISO 9613-1 are very complex with respect to temperature and humidity. This is why Monte-Carlo simulations will have to be used in Section 3.1.3.A to analyze the influence of temperature and humidity variations on the air attenuation coefficient and its effect on the uncertainty of the room absorption coefficient. The procedure for carrying out such Monte-Carlo simulations is described in [130, Part 3, Supplement 1].

D Reverberation Time

The susceptibility of the room absorption coefficient to variations in the reverberation time

$$\frac{\partial \overline{\alpha}}{\partial T} = -\frac{4 \, V_{\text{room}}}{S_{\text{room}}} \cdot \frac{6 \ln(10)}{c \cdot T^2} = -\frac{1}{T} \cdot (\overline{\alpha} + \alpha_{\text{air}}) \tag{3.7}$$

rises with the total absorption in the room $\alpha_{\text{tot}} = \overline{\alpha} + \alpha_{\text{air}}$ and at the same time with the inverse of the reverberation time. Both effects are correlated as high absorption will always lead to short reverberation times, and it could be concluded that the sensitivity to deviations in the reverberation time is important only for higher frequencies as the absorption usually increases with frequency. However, it will be shown in Section 3.1.3.B, that the variation of reverberation times is highest for lower frequencies and thus counterbalances the behavior of the sensitivity coefficient.

E Eyring vs. Sabine Equation

It was already mentioned in Section 2.5 that there is a clear relationship between Eyring and Sabine absorption coefficients and the same holds for the sensitivity coefficients. By calculating the partial derivative of Eq. (2.101), the relationship between the sensitivity coefficients for an input quantity x calculated according to the Eyring and Sabine equation is obtained:

$$\frac{\partial \overline{\alpha}}{\partial x}^{(\text{Eyr})} = \left(1 - \overline{\alpha}^{\text{Eyr}}\right) \cdot \frac{\partial \overline{\alpha}}{\partial x}^{(\text{Sab})} . \tag{3.8}$$

This equation can be used to convert between different conventions, however as was already mentioned before, the standardized methods prescribe the use of the Sabine equation.

3.1.2 Systematic Deviations

The partial derivatives calculated in the previous subsection can now be used to perform the uncertainty analysis with respect to systematic deviations of some of the input quantities. In this context only the quantities related to the room dimensions will be considered, as the other input quantities are usually subject to random variations, not systematic ones.

Additionally, the difference of the absorption coefficient calculated according to the Eyring or Sabine equation will be investigated. Although this aspect is actually not related to the input quantities, it nonetheless is an important systematic deviation and will hence be considered here.

A Room Volume and Surface Area

Using Eq. (2.110) for a single input quantity, it follows that the relative deviation in the output is equal to the relative deviation of the input for a systematic error in the geometrical quantities:

$$\left.\frac{\Delta\overline{\alpha}}{\overline{\alpha}}\right|_{V_{\text{room}}} = \frac{\frac{\partial\overline{\alpha}}{\partial V_{\text{room}}}\cdot\Delta V_{\text{room}}}{\overline{\alpha}} = \frac{\Delta V_{\text{room}}}{V_{\text{room}}}, \tag{3.9}$$

$$\left.\frac{\Delta\overline{\alpha}}{\overline{\alpha}}\right|_{S_{\text{room}}} = \frac{\frac{\partial\overline{\alpha}}{\partial S_{\text{room}}}\cdot\Delta S_{\text{room}}}{\overline{\alpha}} = -\frac{\Delta S_{\text{room}}}{S_{\text{room}}}. \tag{3.10}$$

It should be noted that of course the room volume and surface area are always connected through the dimensions of the room, even if the determination of each measure is probably carried out independently. In any case, it can be assumed that the effect of an incorrect estimation of the room dimensions is negligible, as such errors will probably lie below one percent[1]. Nevertheless, the analysis concerning the sample absorption coefficient and the scattering coefficient will be carried out to investigate whether the error could become larger in those cases.

B Eyring vs. Sabine Equation

As already mentioned before, the room absorption coefficient is usually calculated according to the Sabine equation (Eq. (2.100)), which is a requirement of the measurement standards. The difference between the

---

[1]With the application of room acoustics in mind, relative errors of the absorption coefficient below four percent can be considered low enough to result in errors of the predicted reverberation time less than the *just noticeable difference* (JND), see [46, Section C2]

(more precise) Eyring equation and the Sabine equation concerning the room absorption coefficient can be analytically determined with the help of Eq. (2.101). Considering the Sabine absorption coefficient to be the accepted value, and denoting it as $\overline{\alpha}$, the relative deviation of using the Eyring coefficient α^{Eyr} can be expressed as

$$\left.\frac{\Delta\overline{\alpha}}{\overline{\alpha}}\right|_{\text{Eyr}} = \frac{\overline{\alpha} - \overline{\alpha}^{(\text{Eyr})}}{\overline{\alpha}} = 1 - \frac{1 - e^{-\overline{\alpha}}}{\overline{\alpha}}. \tag{3.11}$$

In Figure 3.1 this relative error is graphed in percent as a function of $\overline{\alpha}$. Typical values of the room absorption coefficient lie below 0.1, so that the error is usually less than 5%. Although this is not a high value, it will be shown in Section 3.2.2.C and Section 3.3.2.C that for the sample absorption coefficient and especially the scattering coefficient the errors can be significantly higher. It can be seen that the error is always positive which shows that the absorption coefficient according to Eyring is always less than the one according to Sabine.

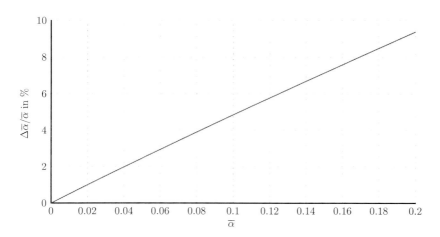

Figure 3.1: Relative error according to Eq. (3.11) in percent between the average room absorption coefficient according to the Eyring and Sabine equation

3.1.3 Random Deviations

The quantities subject to random variations during multiple measurements in the reverberation chamber are the reverberation times and the climatic conditions, the latter especially affecting the air attenuation coefficient. The effect of temperature changes on the speed of sound has already been analyzed and deemed negligible in Section 3.1.1.B.

A Air Attenuation Coefficient

The climatic conditions in the reverberation chamber are only known with finite precision and they may vary, especially if measurements are performed over an extended period of time. A spatial variation of the climatic conditions may also occur but usually the measurements are performed after enough waiting time has passed to let the room settle in order to ensure a homogeneous distribution of temperature and humidity without further air flow. For the application in this context the spatial variation is thus not further considered.

Typically, a correlation exists between air temperature $\Delta\Theta$ and the relative humidity ϕ. This relates to the absolute amount of water vapor in the air, which — when considered constant — determines the fixed relationship between temperature and relative humidity. The temporal changes considered here, however, are related to an exchange of air between the outside and inside when the door of the chamber is opened to either move the source or the microphone(s) to a different location or for moving the absorptive sample into or out of the room. This exchange of air leads to a change also in the absolute humidity and hence the two climatic variables are considered as uncorrelated.

With the assumption of no correlation between $\Delta\Theta$ and ϕ, the influence of each variable on the air attenuation can be calculated and the final result is the superposition according to Eq. (2.115). This has been done using Monte-Carlo simulations for a relative uncertainty of the input parameters between $0\,\%$ and $10\,\%$ and for the third-octave band center frequencies from $2000\,\text{Hz}$ to $5000\,\text{Hz}$, where the influence of air attenuation is high. The Monte-Carlo simulations were performed by generating 10^6 normally distributed random numbers for each value of the relative uncertainty and each input parameter. The uncertainty of the air attenuation coefficient u_m was evaluated for each input parameter and the combined uncertainty was then calculated by:

$$u_m = \sqrt{\left(u_m|_{\Delta\Theta}\right)^2 + \left(u_m|_{\phi}\right)^2}. \tag{3.12}$$

The results are shown as a function of the relative uncertainty of the input parameter in Figure 3.2a and Figure 3.2b for the temperature and relative humidity, respectively, with the third-octave band center frequency as the curve parameter. The combined uncertainty of the air attenuation coefficient at the maximum frequency of $f_c = 5000\,\mathrm{Hz}$ is depicted in Figure 3.2c as a function of the relative uncertainty of the temperature with the relative uncertainty of the relative humidity as the curve parameter.

By comparing Figure 3.2a and Figure 3.2b it can be seen that the relative humidity has a slightly higher influence on the air attenuation coefficient. For both temperature and relative humidity the uncertainty of the air attenuation coefficient is linearly proportional to the relative uncertainty of the respective climate factor. The data also shows that the uncertainty increases with frequency, giving the largest values for the highest frequency considered.

The maximum value of the combined uncertainty is approximately $1\,\mathrm{km}^{-1}$ but one has to keep in mind that the relative uncertainties of the input parameters are rarely as high as $10\,\%$. For typical commercially available devices, the measurement uncertainty given by the manufacturer is $u_{\Delta\Theta} = 0.4\,^\circ\mathrm{C}$ for temperature and $u_\phi = 4.5\,\%$ for relative humidity [135], giving relative uncertainties of $2\,\%$ and $7.5\,\%$, respectively, at the reference climatic conditions. The temporal variations will in most cases not be much higher than this, unless large time spans lie between the measurements.

Finally, the influence of an uncertainty of the climatic conditions on the room absorption coefficient $\overline{\alpha}$ can be determined with the help of Eq. (3.6) as[2]

$$u_{\overline{\alpha}}\big|_m = \frac{4\,V_{\mathrm{room}}}{S_{\mathrm{room}}} \cdot u_m \, , \tag{3.13}$$

which shows that the influence of variations in the climatic conditions increases with the mean free path and hence with the room volume. For typical values of the mean free path in reverberation chambers between $2\,\mathrm{m}$ and $5\,\mathrm{m}$, this leads to a maximum uncertainty of the room absorption coefficient of 0.005. Further investigations on this effect will be carried out in relation to the sample absorption and scattering coefficient in Section 3.2.3.A and Section 3.3.3.A, respectively.

---

[2]By definition, the uncertainty due to random deviations of the input variables is positive, hence the negative sign has been omitted

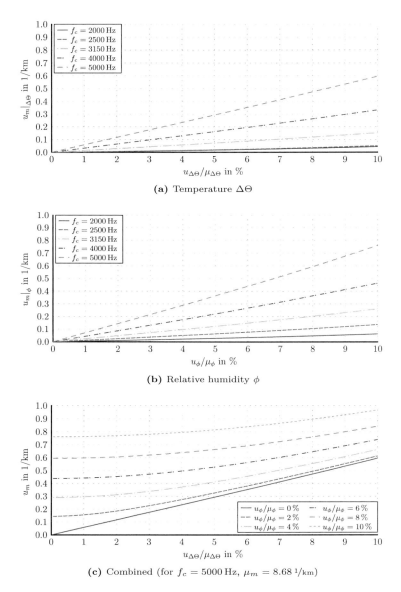

(a) Temperature $\Delta\Theta$

(b) Relative humidity ϕ

(c) Combined (for $f_c = 5000$ Hz, $\mu_m = 8.68\ ^1/\mathrm{km}$)

Figure 3.2: Uncertainty of the air attenuation coefficient as a function of the relative uncertainty of the input parameters $\Delta\Theta$ and ϕ, calculated for average values of $\mu_{\Delta\Theta} = 20\,^\circ\mathrm{C}$ and $\mu_\phi = 60\,\%$

B Reverberation Time

The condition of a perfectly diffuse sound field for the theoretical equations in Section 2.5 to be applicable depends on the Schroeder frequency. Below and close to this frequency, the sound field, and hence the decay rate, may vary significantly with the location of the receiver in the room. In order to overcome this problem, the standards usually require a measurement of the reverberation time for a total of M independent combinations of source and receiver positions — typically $M = 12$ — and use the average value for the calculation of the room absorption coefficient.

Concerning an uncertainty analysis, the spatial standard deviation of the reverberation times according to Eq. (2.113) is of interest[3]. For the application of uncertainty calculations, the actual measurement data can of course be used; however, for an analytical analysis of uncertainties related to the variation of reverberation times a prediction of the standard deviation (or standard error) is necessary. This prediction can be established employing the theory of the spatial variation of decay rates established by DAVY, DUNN, and DUBOUT [136] and DAVY [137]. They showed that the spatial variation of the reverberation time can be estimated analytically, and they developed formulas to predict the standard deviation dependent on the reverberation time T, the dynamic range D of the evaluation of the decay curves and the statistical bandwidth B_1 of the analysis filters.

The statistical filter bandwidth B_1 can be approximated by the nominal (or design) bandwidth with respect to octaves BW (e.g. $BW = 1/3$ for third-octaves) and the corresponding center frequency f_c [138]:

$$B_1(BW, f_c, n) = \frac{\pi}{(2\,n - 1) \cdot \sin(\pi/(2\,n))} \cdot \left(2^{BW/2} - 2^{-BW/2}\right) \cdot f_c . \quad (3.14)$$

For the 5th-order filters used in this study, which are generated in MATLAB and comply with IEC 61260 [139], this becomes

$$B_1(BW, f_c, n = 5) = 1.13 \cdot \left(2^{BW/2} - 2^{-BW/2}\right) \cdot f_c . \quad (3.15)$$

According to the latest paper by DAVY [140] the spatial standard deviation of the reverberation time σ_T can be calculated by

$$\sigma_T(D, BW, f_c, n = 5) = \frac{10}{D^{3/2} \ln(10)} \cdot \sqrt{\frac{720 \cdot F\left(\frac{\ln(10)}{10}\,D\right)}{1.13 \cdot (2^{BW/2} - 2^{-BW/2})}} \cdot \sqrt{\frac{T}{f_c}} , \quad (3.16)$$

---

[3]As mentioned in Table 3.1 other causes for an uncertainty of the reverberation times have not been investigated in this work

with

$$F(x) = 1 - 3 \cdot \frac{1 + e^{-x}}{x} - 12 \cdot \frac{e^{-x}}{x^2} + 12 \cdot \frac{1 - e^{-x}}{x^3} . \qquad (3.17)$$

These equations show that the variation of reverberation times decreases with increasing frequency, as mentioned before. Additionally, it decreases with a larger dynamic range of the decay evaluation D and with increasing filter bandwidth BW.

As an example, Figure 3.3 shows the factor related to $\sqrt{T/f_c}$ in Eq. (3.16) as a function of D in dB and for third-octave ($BW = 1/3$) and octave ($BW = 1$) band filters. The data indicates that a larger bandwidth for evaluation of reverberation times would be beneficial. This would, however, limit the frequency resolution. In any case the dynamic range of evaluation of the decay curves should be as large as the measurement data permits.

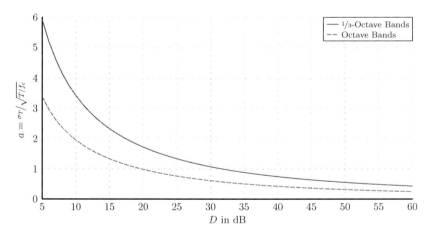

Figure 3.3: The factor $a = \sigma_T / \sqrt{T/f_c}$ (see Eq. (3.16)) as a function of the dynamic range of decay evaluation D in dB for third-octave and octave band filters

For the typical bandwidth of third-octaves ($BW = 1/3$) and commonly used values of the dynamic range of $D = 15\,\text{dB}$, $D = 20\,\text{dB}$ and $D = 30\,\text{dB}$, the uncertainty of the reverberation time obtained for M independent source-receiver combinations can be approximated by

$$u_{T15,1/3} \approx 2.334 \cdot \sqrt{\frac{T}{M \cdot f_c}},$$

$$u_{T20,1/3} \approx 1.724 \cdot \sqrt{\frac{T}{M \cdot f_c}},$$

$$u_{T30,1/3} \approx 1.075 \cdot \sqrt{\frac{T}{M \cdot f_c}}.$$

These expressions can be used to calculate the contribution by the spatial variation of reverberation times to the overall uncertainty in the room absorption coefficient using Eq. (2.114):

$$u_{\overline{\alpha}}\big|_T = \frac{\partial \overline{\alpha}}{\partial T} u_T = \frac{24 \, \log(10) \, V_{\text{room}}}{c \, S_{\text{room}} \, T^2} \cdot u_T. \tag{3.18}$$

It will later turn out that this variation is the main cause of uncertainties in the determination of absorption and scattering coefficients. At this point there will be no further discussion of the result. The analysis will be carried out in connection to the sample absorption and scattering coefficient in Section 3.2.3.B and Section 3.3.3.B, respectively.

The equations for the prediction of spatial standard deviations summarized here are based on the assumption that the theory of statistical room acoustics can be applied, i. e. that frequencies well above the Schroeder frequency are considered. For lower frequencies the low modal overlap usually results in larger variation than is predicted by Eq. (3.16). This problem was addressed first in [136] and a correction factor was suggested in [140].

The suggested factor for the corrected uncertainty u'_T is of the form

$$\frac{u'_T}{u_T} = \sqrt{b_0 + \frac{b_1}{M_s(T)}}, \tag{3.19}$$

where M_s is the statistical modal overlap [141, Appendix C]

$$M_s(T) = \frac{3 \ln(10)}{T} \cdot \frac{\partial N_f}{\partial f}, \tag{3.20}$$

with $\frac{\partial N_f}{\partial f}$ as the modal density (see Eq. (2.90)). The coefficients b_0 and b_1 in Eq. (3.19) have to be determined experimentally through linear regression of the ratio of the experimental to the theoretical standard deviation according to Eq. (3.16) against $1/M_s$. The values obtained by Davy are $b_0 = 0.75$ and $b_1 = 2.78$. However, the data of other studies [11] showed that the values are probably dependent on some other parameter. This will be investigated further in Section 3.2.3.B.

3.1.4 Influence of Higher-Order Derivatives

It was mentioned in Section 2.7 that the propagation of uncertainty is based on a Taylor-series expansion, which is usually stopped after the linear term. However, in some cases a higher-order approximation might be necessary.

With regard to the analysis carried out in this section for the room absorption coefficient, most of the higher-order derivatives are either zero — as for the room volume and the air attenuation coefficient — or their values are much lower than for the first order — as is the case for the room surface area and the speed of sound — so that they can be neglected.

Hence, the only higher-order derivative worth considering is the one related to the reverberation time, which can generally be expressed as a function of the order n as

$$\frac{\partial^n \overline{\alpha}}{\partial T^n} = \frac{24 \ln(10) \, V_{\text{room}}}{c \, S_{\text{room}}} \cdot \frac{(-1)^n \, n!}{T^{n+1}} \cdot \qquad (3.21)$$

To determine the relative influence of the higher-order ($n \geq 2$) terms of the uncertainty $u_{\overline{\alpha}}^{(n)}$, the contributions by each order can be set in relation to the contribution of the linear term (compare Eq. (2.109), Eq. (3.16) and Eq. (3.18)):

$$\left. \frac{u_{\overline{\alpha}}^{(n)}}{u_{\overline{\alpha}}} \right|_T = \frac{\frac{1}{n!} \cdot \frac{\partial^n \overline{\alpha}}{\partial T^n} \cdot u_T^n}{\frac{\partial \overline{\alpha}}{\partial T} \cdot u_T} = u_T^{n-1} \cdot \frac{(-1)^{n-1}}{T^{n-1}},$$

$$= \left(-\frac{a}{\sqrt{M \cdot f_c \cdot T}} \right)^{n-1}, \qquad (3.22)$$

with a as the constant factor depending on the value of D as plotted in Figure 3.3. For the frequency range used in ISO 354 and typical values of M and T, the influence of the order $n = 2$ is at most $3\,\%$ of that of the first order ($n = 1$) and the relative influence of the next-higher order ($n = 3$) is less than $0.1\,\%$. This gives confidence in neglecting the higher-order terms for the rest of the uncertainty analysis.

3.2 Sample Absorption Coefficient (ISO 354)

The analysis carried out in the last section can directly be applied to the measurements according to ISO 354 as the room absorption coefficient is measured under two conditions to obtain the sample absorption coefficient α_s: $\overline{\alpha}_1$ is measured for the empty chamber and $\overline{\alpha}_2$ is measured for the chamber with a sample of surface area S_{sample} inside. The sample absorption coefficient is then calculated according to Eq. (2.102).

3.2.1 Sensitivity Coefficients

It follows from Eq. (2.102) that the sample absorption coefficient does not depend on the room surface area and hence Eq. (3.2) is no longer needed. Instead a new input quantity — the sample surface area S_{sample} — leads to an additional sensitivity coefficient:

$$\frac{\partial \alpha_s}{\partial S_{\text{sample}}} = -\frac{\alpha_s}{S_{\text{sample}}} , \tag{3.23}$$

which shows that the influence of an uncertainty in the sample surface area increases with the amount of absorption of the sample but it decreases with the size of the sample.

Applying the chain rule, the sensitivity coefficients for the influence of an input quantity x on the sample absorption coefficient can be determined:

$$\frac{\partial \alpha_s}{\partial x} = \frac{\partial \alpha_s}{\partial \overline{\alpha}_{1,2}} \cdot \frac{\partial \overline{\alpha}_{1,2}}{\partial x} , \tag{3.24}$$

with

$$\frac{\partial \alpha_s}{\partial \overline{\alpha}_{1,2}} = \mp \frac{S_{\text{room}}}{S_{\text{sample}}} . \tag{3.25}$$

For the sake of brevity, the sensitivity coefficients will not all be repeated for the sample absorption coefficient as hardly any new information can be obtained in doing so. Instead the analysis concerning systematic and random deviations is presented for the individual input quantities.

3.2.2 Systematic Deviations

Regarding the sample absorption coefficient, the cause of systematic deviations can either be the room and sample dimensions or the way of calculating the average room absorption coefficients from the reverberation times (compare Table 3.1). These possible sources of error will be investigated in this section.

Using Eq. (2.110) together with Eq. (3.24) and Eq. (3.25), the systematic deviation of the sample absorption coefficient can be calculated from the deviations of the input quantities as

$$\Delta\alpha_s = \frac{\partial\alpha_s}{\partial\overline{\alpha}_1} \cdot \Delta\overline{\alpha}_1 + \frac{\partial\alpha_s}{\partial\overline{\alpha}_2} \cdot \Delta\overline{\alpha}_2 \,,$$

$$= \frac{S_{\text{room}}}{S_{\text{sample}}} \cdot (\Delta\overline{\alpha}_2 - \Delta\overline{\alpha}_1) \,. \tag{3.26}$$

A Room Volume and Sample Surface Area

The deviation of the sample absorption coefficient with respect to systematic deviations of the room volume is

$$\Delta\alpha_s|_{V_{\text{room}}} = \frac{S_{\text{room}}}{S_{\text{sample}}} \cdot \left(\frac{\overline{\alpha}_2}{V_{\text{room}}} - \frac{\overline{\alpha}_1}{V_{\text{room}}} \right) \cdot \Delta V_{\text{room}} = \alpha_s \cdot \frac{\Delta V_{\text{room}}}{V_{\text{room}}} \,, \tag{3.27}$$

and hence

$$\frac{\Delta\alpha_s}{\alpha_s}\bigg|_{V_{\text{room}}} = \frac{\Delta V_{\text{room}}}{V_{\text{room}}} \,, \tag{3.28}$$

which is the same result as for the average room absorption coefficient (compare Eq. (3.9)) and thus the influence can be deemed negligible with the same reasoning.

For the sample surface area, a similar result is obtained:

$$\frac{\Delta\alpha_s}{\alpha_s}\bigg|_{S_{\text{sample}}} = -\frac{\Delta S_{\text{sample}}}{S_{\text{sample}}} \,, \tag{3.29}$$

with the difference that the relative error in the estimation of the sample surface area could potentially be greater than for the room volume as the sample area is much smaller. It is easy to see that an overestimation of the sample surface area leads to an underestimated sample absorption coefficient.

B Room Surface Area Covered by Sample

The systematic effect on the sample absorption coefficient by neglecting the covered part of the room surface area — i.e. the difference between calculations according to ASTM C423 and ISO 354 — is

$$\Delta\alpha_s|_{S_{\text{covered}}} = \alpha_{s,\text{ASTM}} - \alpha_{s,\text{ISO}} = \overline{\alpha}_1 \,. \tag{3.30}$$

The effect is hence related to the absorption properties of the empty reverberation chamber, with typically $\overline{\alpha}_1 \leq 0.1$. The results without a compensation for this effect will hence always be slightly lower. The reference calculations in this thesis are all based on ISO 354 and so the results of the validation measurements will not be compensated for the covered room surface area.

C Eyring vs. Sabine Equation

The systematic deviation introduced by using Sabine's instead of Eyring's equation in determining the sample absorption coefficient is obtained following Eq. (3.11) and using Eq. (3.26), which yields after some simplification

$$\frac{\Delta \alpha_s}{\alpha_s}\bigg|_{\text{Eyr}} = 1 - \frac{\alpha_s^{(\text{Eyr})}}{\alpha_s} = 1 + \frac{e^{-\overline{\alpha}_2} - e^{-\overline{\alpha}_1}}{\overline{\alpha}_2 - \overline{\alpha}_1}, \qquad (3.31)$$

with the sample absorption coefficient $\alpha_s \equiv \alpha_s^{(\text{Sab})}$ according to Sabine as the reference.

The analysis of this effect is best carried out depending on the room parameter $\overline{\alpha}_1$ and the sample parameter α_s, which is connected to the average room absorption coefficient $\overline{\alpha}_2$ through (compare Eq. (2.102))

$$\overline{\alpha}_2 = \overline{\alpha}_1 + \frac{S_{\text{sample}}}{S_{\text{room}}} \cdot \alpha_s . \qquad (3.32)$$

The relative error according to Eq. (3.31) is plotted in Figure 3.4 in percent as a function of $\overline{\alpha}_1$ for different values of α_s and for a ratio $S_{\text{sample}}/S_{\text{room}} = 1/20$, representing a typical value, e. g. $S_{\text{room}} = 220\,\text{m}^2$ and $S_{\text{sample}} = 11\,\text{m}^2$.

In comparison to Figure 3.1 the error is approximately twice as high, with values for typical rooms of at most 10 %. The dependency on the sample absorption coefficient is relatively low, resulting in a difference between low and fully absorptive samples of roughly two percent. The data shows that although the deviation between the two conventions for calculating $\overline{\alpha}$ is relatively low, the error becomes larger for the calculation of the sample absorption coefficient. It will be shown in Section 3.3.2.C that this becomes even more important for the scattering coefficient.

The (relative) error according to Eq. (3.31) is always positive, indicating that the values of α_s based on calculations on the Eyring equation are always lower than those based on the Sabine equation. This could be one of the causes of sample absorption coefficient measurements returning values greater than one. It has been confirmed by HODGSON [142] that Eyring's formula is more accurate for surfaces with high absorption.

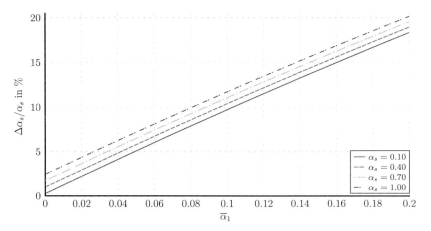

Figure 3.4: Relative error according to Eq. (3.31) in percent between the sample absorption coefficient according to the Eyring and Sabine equation as a function of $\overline{\alpha}_1$ for different values of α_s

3.2.3 Random Deviations

The input quantities subject to random deviations with respect to the sample absorption coefficient are the reverberation time and the climatic conditions influencing the air attenuation coefficient. The findings in Section 3.1.3 will be used and extended in this section.

Applying Eq. (2.115) to Eq. (2.102) results in an equation for the uncertainty of the sample absorption coefficient:

$$
\begin{aligned}
u_{\alpha_s} &= \sqrt{\left(\frac{\partial \alpha_s}{\partial \overline{\alpha}_1} \cdot u_{\overline{\alpha}_1}\right)^2 + \left(\frac{\partial \alpha_s}{\partial \overline{\alpha}_2} \cdot u_{\overline{\alpha}_2}\right)^2}, \\
&= \frac{S_{\text{room}}}{S_{\text{sample}}} \cdot \sqrt{u_{\overline{\alpha}_1}^2 + u_{\overline{\alpha}_2}^2}.
\end{aligned}
\tag{3.33}
$$

To compress the notation in some of the following equations the room volume, the sample surface area and the speed of sound in the empty chamber c_1 will be combined into the room setup constant K in seconds as

$$
K = \frac{24 \ln(10) V_{\text{room}}}{c_1 \cdot S_{\text{sample}}},
\tag{3.34}
$$

with typical values in the range[4] $1\,\mathrm{s} \leq K \leq 4\,\mathrm{s}$ for rooms of regular size $100\,\mathrm{m}^3 \leq V_{\mathrm{room}} \leq 400\,\mathrm{m}^3$ and sample sizes of $10\,\mathrm{m}^2 \leq S_{\mathrm{sample}} \leq 12\,\mathrm{m}^2$. The value corresponding to the reference volume of $V_0 = 200\,\mathrm{m}^3$ given in ISO 354 is approximately $K = 3\,\mathrm{s}$. Rooms with a lower value of K can thus be considered small and rooms with large K are relatively large.

A Air attenuation Coefficient

Applying Eq. (3.33) to the climatic conditions and the influence on the air attenuation coefficient leads to

$$
\begin{aligned}
u_{\alpha_s}\big|_m &= \frac{S_{\mathrm{room}}}{S_{\mathrm{sample}}} \cdot \sqrt{u_{\overline{\alpha_1}}^2\big|_m + u_{\overline{\alpha_2}}^2\big|_m}\,, \\
&= \frac{4\,V_{\mathrm{room}}}{S_{\mathrm{sample}}} \cdot \sqrt{u_{m_1}^2 + u_{m_2}^2}\,, \\
&= \frac{4\,V_{\mathrm{room}}}{S_{\mathrm{sample}}} \cdot u_{m_1} \cdot \sqrt{1 + \left(\frac{u_{m_2}}{u_{m_1}}\right)^2}\,,
\end{aligned}
\tag{3.35}
$$

where Eq. (3.13) has been used to relate the uncertainty of $\overline{\alpha}$ to the uncertainty in m. Without loss of generality it can be assumed that $u_{m_2} = u_{m_1}$, which simplifies the previous equation to yield

$$
u_{\alpha_s}\big|_m = \frac{\sqrt{32}\,V_{\mathrm{room}}}{S_{\mathrm{sample}}} \cdot u_{m_1}\,.
\tag{3.36}
$$

In relation to the room absorption coefficient (compare Eq. (3.13)), the uncertainty of the sample absorption coefficient is higher by a factor of $\sqrt{2}\,\frac{S_{\mathrm{room}}}{S_{\mathrm{sample}}}$, leading to maximum uncertainty values of approximately 0.14, which is definitely significant, although it has to be stressed again that the high variations of the input quantities associated with such values are not likely to be encountered.

With the same reasoning to neglect correlation between temperature and humidity as in Section 3.1.3.A, it is assumed here that the two air attenuation coefficients m_1 and m_2 are not correlated. A further investigation into such effects is out of the scope of this work.

---

[4]These values assume that the correction of the sample surface area for volumes above $200\,\mathrm{m}^3$ according to ISO 354 has been used; otherwise values for K may be larger

B Reverberation Time

With the constant defined in Eq. (3.34) and applying Eq. (3.33) the uncertainty of the sample absorption coefficient with respect to random variations of the reverberation time is given by

$$
\begin{aligned}
u_{\alpha_s}|_T &= K \cdot \sqrt{\left(\frac{u_{T_1}}{T_1^2}\right)^2 + (c_1/c_2)^2 \cdot \left(\frac{u_{T_2}}{T_2^2}\right)^2}, \\
&\approx K \cdot \frac{u_{T_1}}{T_1^2} \cdot \sqrt{1 + \left(\frac{u_{T_2}}{u_{T_1}} \cdot \frac{T_1^2}{T_2^2}\right)^2},
\end{aligned}
\tag{3.37}
$$

where the assumption $c_1/c_2 \approx 1$ has been made to simplify the analysis. This is justified as there is hardly any variation of the speed of sound due to changes in the room temperature. For an error of one percent in the speed of sound the temperature variation would have to be more than five degrees, which corresponds to a variation of 25 % at the reference temperature of $\Delta\Theta = 20\,°\text{C}$. However, typical variations of the temperature rarely exceed one degree, especially in measurement chambers equipped with air conditioning systems.

For a more general uncertainty analysis the parameters u_{T_1} and u_{T_2} can be approximated with the analytic expression by Davy (see Eq. (3.16)). For a worst-case approximation, the minimum value allowed by ISO 354 of $D = 20\,\text{dB}$ will be used (see [5, Section 7.4]), so that the uncertainty of α_s becomes:

$$
u_{\alpha_s}|_T \approx \frac{1.724}{\sqrt{M\,f_c}} \cdot \frac{K}{\sqrt{T_1^3}} \cdot \sqrt{1 + \left(\frac{T_1}{T_2}\right)^3}.
\tag{3.38}
$$

Eq. (3.38) indicates that the uncertainty decreases with increasing values of T_1. Under the assumption of an equal speed of sound as stated before the inverse of T_2 can be related to the reverberation time of the empty chamber T_1 and the sample absorption coefficient α_s by rearranging Eq. (2.102):

$$
\frac{1}{T_2} = \frac{\alpha_s}{K} + \frac{1}{T_1} + \frac{1}{T_{2,\text{max}}} - \frac{1}{T_{1,\text{max}}},
\tag{3.39}
$$

where the maximum theoretically possible value of the reverberation time to obtain $\overline{\alpha} > 0$ (compare Eq. (2.100)) has been defined as

$$
T_{\text{max}} = \frac{6\,\ln(10)}{c \cdot m},
\tag{3.40}
$$

so that $\frac{T_{\text{max}}}{T} > 1$ is always fulfilled. With respect to the spatial variation of reverberation times low frequencies ($f_c \leq 1000\,\text{Hz}$) are of interest. In that frequency range $T_{2,\text{max}} \approx T_{1,\text{max}}$ and hence the factors related to the (change of) the climatic conditions can be neglected for this particular investigation, which typically results in relative errors of T_2 of less than 0.5 %. With this simplification the relation between T_1 and T_2 becomes

$$\frac{1}{T_2} = \frac{\alpha_s}{K} + \frac{1}{T_1}, \tag{3.41}$$

which can be inserted into Eq. (3.38):

$$u_{\alpha_s}|_T \approx \frac{1.724}{\sqrt{M}\,f_c} \cdot \frac{K}{\sqrt{T_1^3}} \cdot \sqrt{1 + \left(\frac{\alpha_s T_1}{K} + 1\right)^3}. \tag{3.42}$$

If correlation exists between the input quantities T_1 and T_2, i. e. the correlation coefficient $r(T_1, T_2) \neq 0$, then according to Eq. (2.116) the uncertainty of the absorption coefficient due to random variations of the reverberation time can be expressed as

$$u_{\alpha_s}|_T \approx \frac{1.724}{\sqrt{M}\,f_c} \cdot \frac{K}{\sqrt{T_1^3}} \cdot \sqrt{1 + \left(\frac{\alpha_s T_1}{K} + 1\right)^3 - 2 \cdot r(T_1, T_2) \cdot \sqrt{\left(\frac{\alpha_s T_1}{K} + 1\right)^3}}. \tag{3.43}$$

Eq. (3.37) can be corrected accordingly. Eq. (3.43) shows that for a strong positive correlation between the reverberation times, the combined uncertainty of α_s is always lower than the one calculated without correlation. In Section 3.2.3.C it will be investigated through verification measurements whether correlation is significant for the measurement of sample absorption coefficients.

Eq. (3.42) and Eq. (3.43) are relatively simple equations to predict the uncertainty of the sample absorption coefficient as a result of the spatial variation of reverberation times, depending only on the following parameters:

- the quotient of V_{room} and S_{sample}, combined in the constant K,

- the reverberation time T_1 of the empty chamber,

- the sample absorption coefficient α_s,

- the third-octave band center frequency f_c, and

- the number M of source-receiver combinations in the sound field.

To gain insight into the behavior of the uncertainty according to Eq. (3.42), Figure 3.5 presents a contour plot of the expanded ($C = 1.96$) uncertainty $u_{0.95}$ to obtain intervals for α_s at the 95 % confidence level as a function of the reverberation time T_1 and the constant K. The values were calculated for a sample absorption coefficient $\alpha_s = 0.75$ and $M \cdot f_c = 1200\,\text{Hz}$, i.e. $M = 12$ at $f_c = 100\,\text{Hz}$ or $M = 6$ at $f_c = 200\,\text{Hz}$ and so forth. The contour lines were drawn at uncertainty levels between 0.05 and 0.1 in steps of 0.01. Additionally, the lower limits of the reverberation time according to ISO 354 corresponding to the effective absorption area of the empty chamber for frequencies below 1000 Hz (see Table 2.1) are marked by the dotted line.

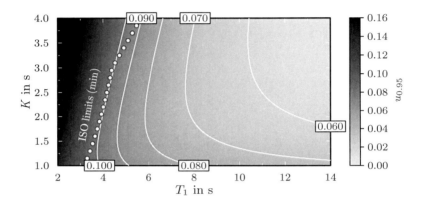

Figure 3.5: Uncertainty of the sample absorption coefficient according to Eq. (3.42) as a function of T_1 and K for a value of $\alpha_s = 0.75$ and $M \cdot f_c = 1200\,\text{Hz}$

The graph should be read as follows: to ensure a maximum uncertainty of $u_{0.95} \leq 0.1$ in a room with $K = 2\,\text{s}$ using $M = 12$ measurement positions at $f_c = 100\,\text{Hz}$, the reverberation time of the empty chamber should be at least $T_1 \geq 4\,\text{s}$ for a sample with $\alpha_s = 0.75$. In the same configuration, a minimum reverberation time of $T_1 \geq 8\,\text{s}$ would be needed to achieve an uncertainty $u_{0.95} \leq 0.07$ and it would be impossible to measure with an uncertainty of less than 0.05 in such a room unless more measurement positions were used to determine the reverberation time. The limits according to ISO 354 lead to uncertainties of approximately 0.1. For large rooms with values of $K > 3\,\text{s}$ the minimum reverberation time allowed by ISO 354 increases and evidently leads to a slightly reduced uncertainty. This of course only holds for an adequately increased sample surface area (see Eq. (2.6.1)).

Before performing a further analysis of Eq. (3.42) regarding the variable parameters, it will first be shown through verification measurements that the equations developed so far are able to correctly describe the behavior of the uncertainty for measurements of the sample absorption coefficient.

C Experimental Verification

The verification measurements presented in this section were all carried out in the reverberation chamber of the Institute of Technical Acoustics (ITA) in Aachen (Figure 3.6a). The chamber has a room volume of $V_{\mathrm{room}} = 123\,\mathrm{m}^3$ and a surface area of $S_{\mathrm{room}} = 178\,\mathrm{m}^2$ and complies with all of the conditions set in ISO 354, although the volume is slightly too small. Curved elements made of plastic are hung inside the chamber to improve diffuseness. The number and location have been determined according to the procedure described in Appendix A of ISO 354.

The chamber is not equipped with an air conditioning system but has a stationary sensor for the measurement of temperature and humidity. The typical relative variation of the climatic conditions during absorption measurements in the reverberation chamber at the ITA is less than 0.5 % for both temperature and relativity humidity. With regard to the findings concerning the influence of the climatic conditions presented in Section 3.2.3.A this leads to the conclusion that in most cases the uncertainty of the climatic conditions does not play a significant role[5].

Figure 3.6b presents the room absorption coefficient of the empty chamber together with the contribution by air absorption and with the limits for the maximum absorption area according to ISO 354 (see Table 2.1) divided by the room surface area. It can be seen that the room absorption coefficient lies below the limits given by the ISO standard for all frequencies.

The data used for verifying the developed equations was taken from previously conducted absorption measurements strictly following the procedure set forth by ISO 354. The two samples used for this study each had a surface area of $S_{\mathrm{sample}} = 10.8\,\mathrm{m}^2$, resulting in a value of $K = 1.83\,\mathrm{s}$.

The reverberation times were obtained from impulse responses measured for a total of $M = 12$ independent source-receiver combinations and evaluated at the third-octave band frequencies between $100\,\mathrm{Hz}$ and $5000\,\mathrm{Hz}$ from the energy decay curves with the minimum dynamic range allowed by ISO 354 of $D = 20\,\mathrm{dB}$.

In Figure 3.7 the average results of the measurements for the two samples mentioned before are depicted. The average reverberation time of the empty chamber and with each of the samples in it is shown in Figure 3.7a and in Figure 3.7b the resulting sample absorption coefficients are plotted.

---

[5] Using the data in Figure 3.2c and the values given here in Eq. (3.36) results in a maximum value of $u_{\alpha s}|_m = 0.0064$, which is certainly negligible.

(a) ITA reverberation chamber

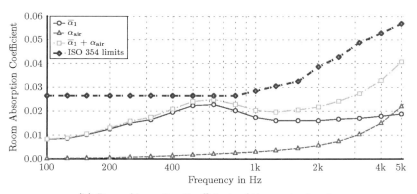

(b) Room Absorption Coefficients of the empty chamber

Figure 3.6: Reverberation chamber at the ITA: (a) Setup with an absorptive sample; (b) Room and air absorption coefficient of the empty chamber with the limits given by ISO 354

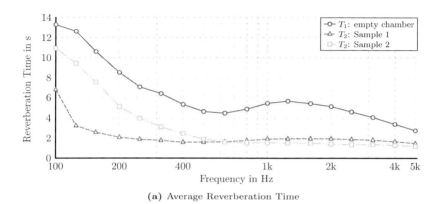

(a) Average Reverberation Time

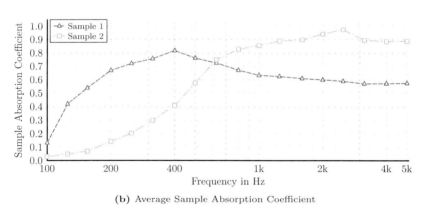

(b) Average Sample Absorption Coefficient

Figure 3.7: Average results of (a) the reverberation time and (b) the sample absorption coefficient for the two samples used in the verification measurements for the uncertainty of absorption coefficients

It was mentioned in Section 2.7.2 that the equations for uncertainty propagation have to be modified if correlation exists between the input quantities. To check for this effect in the case of absorption measurements, the cross correlation $r(T_1, T_2)$ between the reverberation times T_1 and T_2 has been calculated across frequency using the MATLAB function `corrcoef`. A value of the correlation coefficient of plus or minus one would indicate perfect (anti-)correlation, whereas for completely uncorrelated variables the correlation coefficient would be zero. The result is depicted in Figure 3.8 for both of the samples used here.

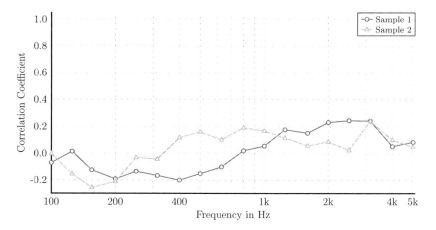

Figure 3.8: Correlation coefficient between the reverberation times obtained during sample absorption measurements for the two samples used in this study

As can be observed, the absolute values of the correlation coefficient do not exceed 0.25, which can be considered low enough to neglect it. To verify this, the `corrcoef` function was used to calculate the probability p testing the hypothesis of no correlation. It was found for the data presented here that the correlation coefficients are not significantly different from zero at the $p = 0.05$ level. The explanation is that once the absorptive sample is inserted into the chamber the sound field is changed enough so that the results of the reverberation time are no longer correlated. It can be argued that for samples that are much less absorptive than the ones used here (i. e. $\alpha_s \ll 0.1$), the correlation could become significant. However, for the uncertainty analysis in this section samples with higher absorption are of interest and hence correlation does not have to be considered. For the remainder of this section $r(T_1, T_2) = 0$ is assumed and hence Eq. (3.42) will be used.

The measurement data was evaluated concerning the uncertainty due to the spatial variation of reverberation times. For this, the measurement results obtained for the different source-receiver combinations were evaluated using Eq. (2.113), which gives the actual uncertainty of the sample absorption coefficient.

Additionally the evaluation using uncertainty propagation was performed, once with the measured uncertainty of the reverberation time (Eq. (3.37)) and once with the approximation following Davy (Eq. (3.42)).

In the latter case, the correction factor related to the modal density (see Section 3.1.3.B) was also included. The first coefficient in Eq. (3.19) was set to $b_0 = 1$ with the reasoning that for large values of the modal overlap M_s the results without and with the correction should be identical. After an analysis of the coefficient b_1 obtained through linear regression as suggested by DAVY [140], the following equations for the corrected standard uncertainty of the reverberation time u'_T were found as

$$\frac{u'_{T_1}}{u_{T_1}} = \sqrt{1 + \frac{1}{4\,M_s(T_1)}}\,, \tag{3.44}$$

$$\frac{u'_{T_2}}{u_{T_2}} = \sqrt{1 + \frac{1 + T_1 - T_2}{4\,M_s(T_2)}}\,, \tag{3.45}$$

for the reverberation times T_1 and T_2, respectively. Figure 3.9 presents the correction factors calculated for the reverberation times shown in Figure 3.7a as a function of frequency. The data shows that for low frequencies the actual variation of the reverberation time can be up to 2.5 times larger than predicted by Eq. (3.42) and that the effect is increased once the sample is placed inside the measurement chamber. This can be explained by the fact that the assumption of an isotropic sound field is violated by the energy that is directed towards the absorber. Above approximately 400 Hz, which is the Schroeder frequency of the empty reverberation chamber, the influence of the modal correction factor is negligible as the statistical modal overlap is high enough.

The correction factors according to Eq. (3.44) and Eq. (3.45) can be inserted into Eq. (3.42) to yield the corrected uncertainty of the sample absorption coefficient

$$u'_{\alpha_s}\big|_T \approx \frac{1.724}{\sqrt{M\,f_c}} \cdot \frac{K}{\sqrt{T_1^3}} \cdot \sqrt{1 + \frac{1}{4\,M_s(T_1)} + \left(1 + \frac{1 + T_1 - T_2}{4\,M_s(T_2)}\right) \cdot \left(\frac{T_1}{T_2}\right)^3}\,, \tag{3.46}$$

where T_2 can be expressed as a function of T_1 and α_s with the help of Eq. (3.41) as before.

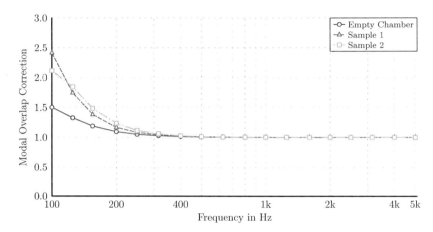

Figure 3.9: Modal correction factors according to Eq. (3.44) and Eq. (3.45) for the reverberation times shown in Figure 3.7a as a function of frequency

It must be stressed that the modal correction factors have been empirically determined for a total of ten different samples but all in the same reverberation chamber. Further investigations with more variation of the measurement setup including chambers of very different size and shape will have to be carried out. Nonetheless, for the remainder of this study the presented correction terms will be used.

The expanded uncertainty with a coverage factor of $C = 1.96$ (see Section 2.7.2) is depicted in Figure 3.10 as a function of frequency for both of the samples used. The reverberation times were evaluated with a dynamic range $D = 20\,\text{dB}$ and $M = 12$ was used to calculate average values and standard uncertainties. There is very good agreement between the uncertainty based on the measured sample absorption coefficients at all source-microphone combinations (solid curve with circle markers) and the one calculated by uncertainty propagation with the measured values of u_{T_1} and u_{T_2} (dashed curve with triangle markers). This suggests that concerning the uncertainty due to spatial variations of the input quantities the reverberation time is the primary factor. It also indicates that the linear approximation in the uncertainty propagation is sufficient and that correlation between the reverberation times can be neglected regarding the sample absorption coefficient.

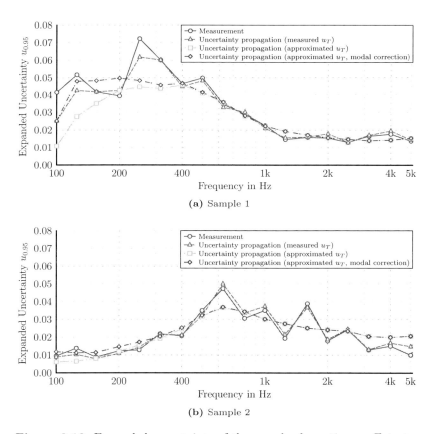

(a) Sample 1

(b) Sample 2

Figure 3.10: Expanded uncertainty of the sample absorption coefficient as a function of frequency evaluated for the two samples used in this study for a value of $D = 20\,\mathrm{dB}$ and $M = 12$

In comparison to the result using the actual reverberation uncertainty data, the approximation according to Eq. (3.16) (dash-dotted curve with square markers) does not follow the actual uncertainty as precisely, as could be expected. However, above 400 Hz there is very good agreement, although at high frequencies the uncertainty is overestimated for Sample 2 (Figure 3.10b). Below 400 Hz the approximated uncertainty without the modal correction underestimates the actual value by a factor of 2–3. The correction factors according to Eq. (3.44) and Eq. (3.45) yield a much better result (dash-dotted curve with diamond markers) even for very low frequencies. The fact that this holds for two samples with very different absorptive properties gives confidence to use the general expression in Eq. (3.46) for a further analysis.

An interesting fact that can be read from the data in Figure 3.10 is that the uncertainty is not necessarily highest for the lowest frequencies, as is the case for the uncertainty of the individual reverberation times. It is rather connected to the absorptive properties of the sample, giving large values even in the mid-frequencies as can be seen for Sample 2 in Figure 3.10b. It can thus be concluded that it is always more problematic to measure highly absorptive samples with a high precision in the low and medium frequency range. This can be confirmed by analyzing Eq. (3.42), which shows that the uncertainty of the sample absorption coefficient is approximately proportional to $\sqrt{\alpha_s^3/f_c}$. In any case, for high frequencies the uncertainty will usually decrease.

D Further Analysis

After having verified that the equations obtained so far produce valid results regarding the uncertainty of the sample absorption coefficient related to the spatial variation of reverberation times, a further analysis of the influencing factors can be carried out in this section. The correction related to the statistical modal overlap has proven to be significant and hence for the following analysis Eq. (3.46) will be used.

Of the factors influencing the uncertainty, the room constant K and the reverberation time T_1 of the empty reverberation chamber are more or less fixed for a certain measurement environment. The frequencies are also set by the standards and hence the variable factors are the sample absorption coefficient α_s and the number of independent source-receiver combinations M, of which only the latter can be freely chosen. The analysis in this section will consist of determining a minimum value for M which would ensure a specified maximum value of the expanded uncertainty $u_{0.95,\max}$ ($C = 1.96$, see Section 2.7.2) for the sample absorption coefficient at a given frequency for an absorber with a given value of α_s at that frequency. In the form of an equation this is represented by (compare Eq. (3.46))

$$M_{\min} \geq \left(\frac{1.96 \cdot 1.724}{u_{0.95,\max}} \right)^2 \cdot \frac{K^2}{f_c \, T_1^3}$$

$$\cdot \left[1 + \frac{1}{4 \, M_s(T_1)} + \left(1 + \frac{1 + T_1 - T_2}{4 \, M_s(T_2)} \right) \cdot \left(\frac{T_1}{T_2} \right)^3 \right] . \qquad (3.47)$$

In analogy to Figure 3.5, the values of $M_{\min}$ will be shown as contour plots for $2\,\mathrm{s} \leq T_1 \leq 16\,\mathrm{s}$ and $1\,\mathrm{s} \leq K \leq 4\,\mathrm{s}$, covering the typical range of values. As a worst-case example, the uncertainty will be investigated at the lowest frequency permitted by ISO 354, which is $f_c = 100\,\mathrm{Hz}$. It will be assumed that the sample has a specified maximum absorption coefficient at that frequency, which will be chosen as one of $\alpha_s = 0.25$, $\alpha_s = 0.5$ or $\alpha_s = 0.75$ to observe the variation connected to this parameter. The maximum uncertainty for the analysis presented here was specified as $u_{0.95,\max} = 0.1$, which is a relatively high value. Nevertheless, the data will show that even for such a high value the necessary number of measurement positions can be substantially higher than the recommendation of $M_{\min} = 12$ in ISO 354.

In Figure 3.11 the results of the minimum number of independent source-receiver combinations according to Eq. (3.47)) are shown as contour plots, calculated for the values mentioned in the previous paragraph. The contour lines are drawn for the values of $M_{\min}$ between 2 and 30 in steps of 2. Additionally the minimum reverberation times allowed by ISO 354 are indicated by the dotted line. The bend in the contours occurs for $K \geq 3\,\mathrm{s}$ due to the ISO correction of the sample size for $V \geq 200\,\mathrm{m}^3$.

The results in the case of low absorptivity ($\alpha_s = 0.25$) of the sample presented in Figure 3.11a show that the minimum number of measurement positions recommended by ISO 354 is surely enough to achieve an extended uncertainty of $u_{0.95,\max} = 0.1$. For all combinations of the room setup and the reverberation time T_1 that fulfill the requirements of the standard the minimum number is even as low as $M_{\min} = 4$ for such a low value of the sample absorption.

In general — and this is true for all the results — more reverberant chambers require less measurement positions to achieve a given uncertainty. This also holds for larger chambers although it has to be noted again that a significantly larger sample is required in that case. For the maximum value of $K = 4\,\mathrm{s}$ the sample surface area should be at least $S_{\mathrm{sample}} = 20\,\mathrm{m}^2$.

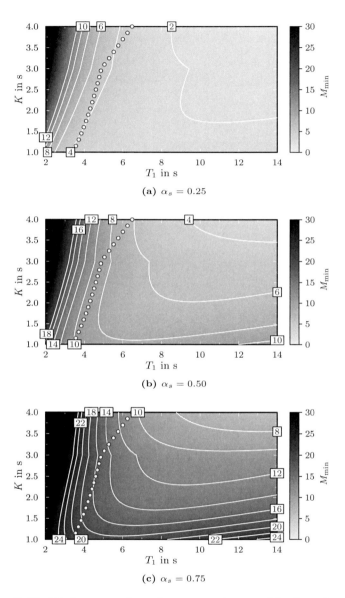

Figure 3.11: Minimum number of source-receiver combinations (according to Eq. (3.47)) needed to achieve $u_{0.95,\mathrm{max}} = 0.1$, calculated at $f_c = 100\,\mathrm{Hz}$ and for different values of α_s

For a medium value of the absorption coefficient ($\alpha_s = 0.50$) the necessary number of measurement positions increases slightly (Figure 3.11b). In chambers that comply with ISO 354 approximately 6–10 independent source-receiver combinations are enough to measure with the given uncertainty. If the sample has an absorption coefficient of $\alpha_s = 0.75$ many more measurement positions are needed as indicated by the results in Figure 3.11c. The recommendation of $M = 12$ in ISO 354 would in that case only yield the specified uncertainty for rooms with $T_1 > 6\,\mathrm{s}$ and $K > 2.5\,\mathrm{s}$.

These results show that the absorptive property of the sample to be measured determines the precision that can be achieved. In most chambers adhering to the ISO guidelines, the suggested value of $M_{\mathrm{min}} = 12$ seems to guarantee an adequate precision. However, for samples that are more absorptive at low frequencies — as it may be the case with resonance absorbers — this is no longer true. It is also questionable whether an expanded uncertainty of 0.1 is low enough, but the answer to this question is not within the scope of this work. For a higher precision, more effort is required for the measurements. It will be shown in 3.3.3 that the situation in the case of scattering coefficient measurements is even more complex.

3.3 Sample Scattering Coefficient (ISO 17497-1)

The measurement of the scattering coefficient in a reverberation chamber is strongly related to the measurement of absorption coefficients (compare Eq. (2.104) and Eq. (2.105)) and hence the results in Section 3.2 can directly be applied in this section.

Due to the more complex equation to calculate the scattering coefficient — also involving two additional measurements — it is likely that the susceptibility to deviations in the input quantities is greater. This is actually often encountered in practical measurement situations. The findings in this section shall hence explain the main sources of this uncertainty.

3.3.1 Sensitivity Coefficients

The sensitivity coefficients regarding the two absorption coefficients α_s and α_{spec}, respectively, are

$$\frac{\partial s}{\partial \alpha_s} = -\frac{1 - s}{1 - \alpha_s}\,, \tag{3.48}$$

and

$$\frac{\partial s}{\partial \alpha_{\mathrm{spec}}} = \frac{1}{1 - \alpha_s}\,. \tag{3.49}$$

91

Additionally, the sensitivity coefficients concerning the room volume and the sample surface area, respectively, are obtained as

$$\frac{\partial s}{\partial V_{\text{room}}} = \frac{1}{1 - \alpha_s} \cdot \frac{s}{V_{\text{room}}}, \tag{3.50}$$

and

$$\frac{\partial s}{\partial S_{\text{sample}}} = -\frac{1}{1 - \alpha_s} \cdot \frac{s}{S_{\text{sample}}}. \tag{3.51}$$

The fact that all of these coefficients strongly depend on the sample absorption coefficient already supports the reasoning in Section 2.6.2 to limit the absorptive properties of the sample. It follows that the sample absorption coefficient should always be as low as possible to keep the influence of uncertainties to a minimum. This will be further explored in the following sections.

3.3.2 Systematic Deviations

In analogy to the previous section, the chain rule will be used to calculate the deviation of the scattering coefficient for systematic deviations of the input quantities. With the help of Eq. (2.110) and Eq. (3.26), the following equation is obtained:

$$\begin{aligned} \Delta s &= \frac{\partial s}{\partial \alpha_s} \cdot \Delta \alpha_s + \frac{\partial s}{\partial \alpha_{\text{spec}}} \cdot \Delta \alpha_{\text{spec}}, \\ &= \frac{1}{1 - \alpha_s} \cdot (\Delta \alpha_{\text{spec}} - (1 - s) \cdot \Delta \alpha_s). \end{aligned} \tag{3.52}$$

A Room Volume and Sample Surface Area

The relative deviation of s due to (relative) systematic errors of the room volume is

$$\frac{\Delta s}{s}\bigg|_{V_{\text{room}}} = \frac{1}{1 - \alpha_s} \cdot \frac{\Delta V_{\text{room}}}{V_{\text{room}}}, \tag{3.53}$$

and — as for the sample absorption coefficient before — a similar result is obtained for the sample surface area:

$$\frac{\Delta s}{s}\bigg|_{S_{\text{sample}}} = -\frac{1}{1 - \alpha_s} \cdot \frac{\Delta S_{\text{sample}}}{S_{\text{sample}}}. \tag{3.54}$$

Compared to Eq. (3.28) and Eq. (3.29) the influence of an incorrectly estimated room volume or sample surface area is increased by the factor

$\frac{1}{1-\alpha_s}$. For the maximum value of $\alpha_s = 0.5$ allowed by ISO 17497-1, this results in deviations that are twice as high as for the sample absorption coefficient, which can be significant.

B Room Surface Area Covered by Sample

The error introduced by neglecting the room surface area covered by the sample of course also influences the scattering coefficient. With Eq. (3.52) and Eq. (3.30) the deviation of the scattering coefficient can be expressed as

$$\Delta s\big|_{S_{\text{covered}}} = \frac{1}{1-\alpha_s} \cdot (\overline{\alpha}_3 - (1-s) \cdot \overline{\alpha}_1) \,,$$

$$= \frac{1}{1-\alpha_s} \cdot \left(s \cdot \overline{\alpha}_1 + \frac{S_{\text{sample}}}{S_{\text{room}}} \cdot s_{\text{base}}\right) \,, \quad (3.55)$$

where typically $\frac{S_{\text{sample}}}{S_{\text{room}}} \cdot s_{\text{base}} \leq 0.01$ holds for setups that comply with ISO 17497-1 (compare Table 2.2) and thus a good approximation is given by

$$\frac{\Delta s}{s}\bigg|_{S_{\text{covered}}} \approx \frac{\overline{\alpha}_1}{1-\alpha_s} \,. \quad (3.56)$$

It follows that — just as for the room volume and sample surface area — the relative deviation caused by neglecting the room surface area covered by the sample is increased by the factor $\frac{1}{1-\alpha_s}$. In comparison to Eq. (3.30), were the absolute error was related to the empty chamber absorption coefficient, in the case of the scattering coefficient it is the relative error, indicating that the influence is greater for higher values of the scattering coefficient.

Figure 3.12 shows the relative error of the scattering coefficient according to Eq. (3.56) in percent as a function of $\overline{\alpha}_1$ for different values of the sample absorption coefficient α_s. The data shows that the sample absorption coefficient has a very high impact on the relative error. In comparison to Figure 3.4, the relative error of the scattering coefficient can be substantially higher, especially for sample absorption coefficients $\alpha_s \geq 0.3$, yielding errors of far more than 15 %. This again gives reason to limit the maximum value of α_s for measurements of the scattering coefficient.

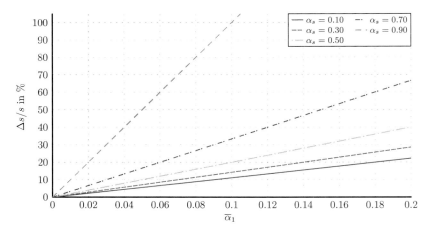

Figure 3.12: Relative error of the scattering coefficient according to Eq. (3.56) due to neglecting the surface area covered by the sample as a function of $\overline{\alpha}_1$ for different values of the sample absorption coefficient α_s

C Eyring vs. Sabine Equation

Calculating the room absorption coefficients from the measured reverberation times can be done by either the Sabine or Eyring equation. This obviously also effects the scattering coefficient because the room absorption in four different conditions has to be determined. As for the sample absorption coefficient in Section 3.2.2.C (see Eq. (3.31)) the value of the scattering coefficient calculated with the $\overline{\alpha}_i$ according to Sabine's equation will be considered as the reference.

The relative deviation due to using Eyring's equation instead is

$$
\left.\frac{\Delta s}{s}\right|_{\mathrm{Eyr}} = 1 - \frac{s^{(\mathrm{Eyr})}}{s},
$$

$$
= 1 + \frac{\mathrm{e}^{-\overline{\alpha}_4} - \mathrm{e}^{-\overline{\alpha}_3} - \mathrm{e}^{-\overline{\alpha}_2} + \mathrm{e}^{-\overline{\alpha}_1}}{\overline{\alpha}_4 - \overline{\alpha}_3 - \overline{\alpha}_2 + \overline{\alpha}_1} \cdot \frac{\frac{S_{\mathrm{sample}}}{S_{\mathrm{room}}} + \overline{\alpha}_1 - \overline{\alpha}_2}{\frac{S_{\mathrm{sample}}}{S_{\mathrm{room}}} + \mathrm{e}^{-\overline{\alpha}_2} - \mathrm{e}^{-\overline{\alpha}_1}}.
$$

$$(3.57)$$

Unfortunately, Eq. (3.57) cannot be brought into a more compact form. It is also impractical to have an expression depending on the individual room absorption coefficients. Hence — similarly to Eq. (3.32) — the absorption coefficients $\overline{\alpha}_3$ and $\overline{\alpha}_4$, respectively, are better expressed by the sample absorption coefficient α_s and the scattering coefficients of the baseplate and the sample (compare Eq. (2.105) and Eq. (2.106)) :

$$\overline{\alpha}_3 = \overline{\alpha}_1 + \frac{S_{\text{sample}}}{S_{\text{room}}} \cdot s_{\text{base}} , \qquad (3.58)$$

$$\overline{\alpha}_4 = \overline{\alpha}_3 + \frac{S_{\text{sample}}}{S_{\text{room}}} \cdot [\alpha_s + s \cdot (1 - \alpha_s)] . \qquad (3.59)$$

The relative deviation according to Eq. (3.57) has been evaluated for different values of α_s, s_{base} and s and the results are shown in Figure 3.13 as a function of the room absorption coefficient $\overline{\alpha}_1$ of the empty reverberation chamber. As in Figure 3.4, a ratio of $S_{\text{sample}}/S_{\text{room}} = 1/20$ was used. The calculations have been carried out for typical values of the baseplate scattering coefficient of $s_{\text{base}} = 0.01$ (Figure 3.13a), $s_{\text{base}} = 0.10$ (Figure 3.13b) and $s_{\text{base}} = 0.20$ (Figure 3.13c). The shaded regions in the graphs correspond to values of the scattering coefficient between zero and one, where the lines have been drawn for the maximum value of s. The data is depicted for values of α_s between 0–0.5.

As for the case of the sample absorption coefficient (Figure 3.4) the relative deviation due to the method of calculating the room absorption coefficient increases proportionally to $\overline{\alpha}_1$. In comparison to the data presented in Figure 3.4 the influence of α_s is much more pronounced for the scattering coefficient as has been stated several times before in this section.

The results for the different values of s_{base} in Figure 3.13a–Figure 3.13c show that the scattering properties of the baseplate do not necessarily lead to an increased deviation but the spread of the results for different values of the scattering coefficient becomes larger, indicated by the larger area of the shaded regions. Interestingly, for an increasing s_{base} the behavior of the results changes with respect to the scattering coefficient.

In Figure 3.13a the maximum relative error is obtained for a maximum value of s, which holds for all sample absorption coefficients. However as the baseplate scattering increases, the relation between the relative error and the scattering coefficient depends on α_s. For low absorption, the maximum deviation occurs for large scattering coefficients, whereas for values of $\alpha_s > 0.2$ this is reversed and large deviations are obtained for low scattering coefficients. This can be observed by the change of the relation between the shaded regions and the curves drawn for the maximum value of s.

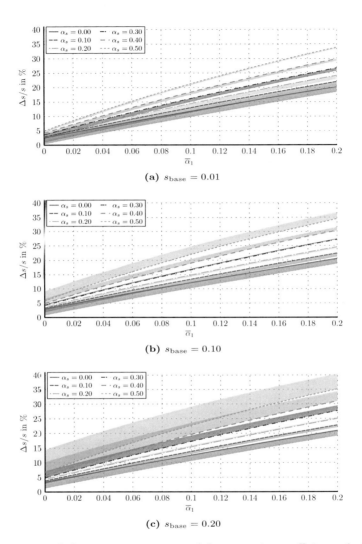

(a) $s_{\mathrm{base}} = 0.01$

(b) $s_{\mathrm{base}} = 0.10$

(c) $s_{\mathrm{base}} = 0.20$

Figure 3.13: Relative error in percent of the scattering coefficient calculated with room absorption coefficients according to Eyring or Sabine; evaluated for different values of the baseplate scattering coefficient and the sample absorption coefficient. The shaded regions correspond to values of the scattering coefficient between zero and one, where the lines have been drawn for the maximum value of s

In general, it can be concluded that the influence of the method of calculating the $\overline{\alpha}_i$ is significantly larger for the scattering coefficient than for the sample absorption coefficient, where relative errors for typical rooms were found to be less than 10 %. Regarding the scattering coefficient, the relative error lies between 5 %–30 % for rooms with $\overline{\alpha}_1 < 0.1$ with a strong dependency on the sample absorption coefficient. It is evident from the analysis presented here that the unnecessary simplification demanded by the standards leads to significant errors that could easily be avoided as the effort of calculating the room absorption according to Eq. (2.99) instead of Eq. (2.100) is not high.

3.3.3 Random Deviations

As in Section 3.2.3, the propagation of uncertainty can be applied to the measurement of the scattering coefficient regarding random errors. This results in

$$
\begin{aligned}
u_s &= \sqrt{\left(\frac{\partial s}{\partial \alpha_s} \cdot u_{\alpha_s}\right)^2 + \left(\frac{\partial s}{\partial \alpha_{\text{spec}}} \cdot u_{\alpha_{\text{spec}}}\right)^2}, \\
&= \frac{1}{1 - \alpha_s} \cdot \sqrt{(1 - s)^2 \cdot u_{\alpha_s}^2 + u_{\alpha_{\text{spec}}}^2}.
\end{aligned}
\tag{3.60}
$$

As for the systematic deviations in Section 3.3.2, the uncertainty depends both on the absorptive as well as the scattering properties of the sample to be measured.

A Air Attenuation Coefficient

Regarding variations of the climatic conditions, the uncertainty of the scattering coefficient is calculated using Eq. (3.60) and Eq. (3.35) as

$$
\begin{aligned}
u_s|_m &= \frac{1}{1 - \alpha_s} \cdot \sqrt{(1 - s)^2 \cdot u_{\alpha_s}^2|_m + u_{\alpha_{\text{spec}}}^2|_m}, \\
&= \frac{4 V_{\text{room}}}{S_{\text{sample}}} \cdot \frac{1}{1 - \alpha_s} \cdot \sqrt{(1 - s)^2 \cdot (u_{m_1}^2 + u_{m_2}^2) + (u_{m_3}^2 + u_{m_4}^2)}, \\
&= \frac{4 V_{\text{room}}}{S_{\text{sample}}} \cdot \frac{u_{m_1}}{1 - \alpha_s} \cdot \sqrt{(1 - s)^2 \cdot \left[1 + \left(\frac{u_{m_2}}{u_{m_1}}\right)^2\right] + \left(\frac{u_{m_3}}{u_{m_1}}\right)^2 + \left(\frac{u_{m_4}}{u_{m_1}}\right)^2},
\end{aligned}
\tag{3.61}
$$

and with the reasonable assumption that $u_{m_4} = u_{m_3} = u_{m_2} = u_{m_1}$ (see Section 3.2.3.A):

$$u_s|_m = \frac{\sqrt{32}\, V_{\text{room}}}{S_{\text{sample}}} \cdot u_{m_1} \cdot \frac{\sqrt{(1-s)^2 + 1}}{1 - \alpha_s},$$

$$= u_{\alpha_s}|_m \cdot \frac{\sqrt{(1-s)^2 + 1}}{1 - \alpha_s}. \tag{3.62}$$

Here, the correlation between m_1–m_4 has been neglected as for the sample absorption coefficient in Section 3.2.3.A.

The uncertainty according to Eq. (3.62) depends strongly on the sample absorption and interestingly it reaches a maximum value for a completely specularly reflecting sample ($s = 1$). To get a further impression, the factor connecting the uncertainty of the sample absorption coefficient to that of the scattering coefficient is depicted in Figure 3.14 as a function of α_s for different values of s.

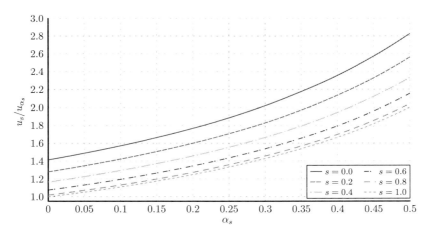

Figure 3.14: Relationship between u_s and u_{α_s} due to variations of the air attenuation coefficient as a function of α_s for different values of the scattering coefficient

For a specularly reflecting sample with an absorption coefficient of 0.5 the uncertainty of the scattering coefficient due to variations of the climatic conditions can be almost three times as high as for the sample absorption coefficient. If the sample scatters all energy, the uncertainty is still twice as high as u_{α_s} for $\alpha_s = 0.5$. In any case, the uncertainty of the scattering coefficient is always larger than for the sample absorption coefficient.

B Reverberation Time

With $u_{\alpha_s}|_T$ and $u_{\alpha_\text{spec}}|_T$ according to Eq. (3.37) for $c_4 = c_3 = c_2 = c_1$, the uncertainty of the scattering coefficient with regard to random variations of the reverberation times is calculated as:

$$
u_s|_T = \frac{K}{1-\alpha_s} \cdot \sqrt{(1-s)^2 \cdot \sum_{i=1}^{2} \frac{u_{T_i}^2}{T_i^4} + \sum_{i=3}^{4} \frac{u_{T_i}^2}{T_i^4}},
$$

$$
= \frac{K}{1-\alpha_s} \cdot \frac{u_{T_1}}{T_1^2}
$$

$$
\cdot \sqrt{(1-s)^2 \cdot \left[1 + \left(\frac{u_{T_2}}{u_{T_1}} \frac{T_1^2}{T_2^2}\right)^2\right] + \left(\frac{u_{T_3}}{u_{T_1}} \frac{T_1^2}{T_3^2}\right)^2 + \left(\frac{u_{T_4}}{u_{T_1}} \frac{T_1^2}{T_4^2}\right)^2}.
$$

$$(3.63)$$

In analogy to Eq. (3.38) the approximation of u_T according to Eq. (3.16) can be used for a general expression of the uncertainty of the scattering coefficient, using $D = 15\,\text{dB}$ as demanded by ISO 17497-1:

$$
u_s|_T \approx \frac{2.334}{\sqrt{M}\,f_c} \cdot \frac{K}{1-\alpha_s} \cdot \frac{1}{\sqrt{T_1^3}} \cdot \sqrt{(1-s)^2 \cdot \left[1 + \left(\frac{T_1}{T_2}\right)^3\right] + \left(\frac{T_1}{T_3}\right)^3 + \left(\frac{T_1}{T_4}\right)^3}.
$$

$$(3.64)$$

As already covered in Section 3.2.3.B, T_2 can be expressed as a function of T_1 and α_s by Eq. (3.41). Similarly, the remaining two reverberation times T_3 and T_4, respectively, can be related to the scattering coefficient of the baseplate and the sample by (compare Eq. (3.58) and Eq. (3.59)):

$$
\frac{1}{T_3} = \frac{s_\text{base}}{K} + \frac{1}{T_1},
$$

$$(3.65)$$

and

$$
\frac{1}{T_4} = \frac{\alpha_s + s \cdot (1-\alpha_s)}{K} + \frac{1}{T_3}.
$$

$$(3.66)$$

In Eq. (3.65) and Eq. (3.66) the effect of (a change of) the climatic conditions has been neglected with the same reasoning as in Eq. (3.41). For the sake of compactness, Eq. (3.64) will not be explicitly given with Eq. (3.65) and Eq. (3.66) inserted.

Nonetheless, as for the sample absorption coefficient in Section 3.2.3.B it can be stated that the uncertainty of the sample scattering coefficient can be expressed as a function of

- the quotient of V_{room} and S_{sample}, combined in the constant K,
- the reverberation time T_1 of the empty chamber,
- the sample absorption coefficient α_s,
- the baseplate scattering coefficient s_{base},
- the sample scattering coefficient s,
- the third-octave band center frequency f_c, and
- the number M of source-receiver combinations in the sound field.

Figure 3.15 presents a contour plot of the uncertainty according to Eq. (3.64) expanded with $C = 1.96$ (see Section 2.7.2) as a function of T_1 and K. The values were calculated for a sample scattering and absorption coefficient, respectively, of $s = 0.5$ and $\alpha_s = 0.25$, a baseplate scattering coefficient $s_{\text{base}} = 0.05$ and $M \cdot f_c = 1200\,\text{Hz}$. The contour lines were drawn at uncertainty levels between 0.10 and 0.20 in steps of 0.02. Additionally, the lower limits of the reverberation time according to ISO 174097-1 (Eq. (2.107)) are indicated by the dotted line.

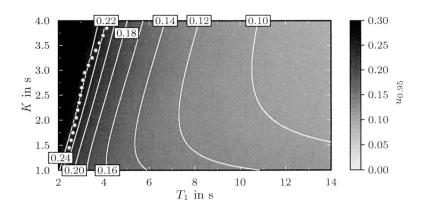

Figure 3.15: Uncertainty of the sample scattering coefficient according to Eq. (3.64) as a function of T_1 and K for a value of $s = 0.5$, $\alpha_s = 0.25$, $s_{\text{base}} = 0.05$ and $M \cdot f_c = 1200\,\text{Hz}$

Compared to the results for the sample absorption coefficient in Figure 3.5 the uncertainty is approximately doubled for the scattering coefficient, indicating how difficult it is to precisely perform such kind of measurements. The behavior in relation to K and T_1 is similar to the case of α_s, yielding large uncertainties for small rooms with short reverberation times.

The limits set forth by the ISO 17497-1 standard, which are significantly below the limits in ISO 354, lead to uncertainties of roughly $u_{0.95} = 0.24$ for the set of parameters used in the example. Further analysis on the uncertainty of the scattering coefficient related to the spatial variation of reverberation times will be carried out after verifying the correctness of the equations by measurements.

For a correlation of the reverberation times the expression for the uncertainty has to be modified, as was mentioned in Section 2.7.2. In the case of the sample absorption coefficient in Section 3.2.3.C it was found that the cross-correlation between T_1 and T_2 was negligible in most cases, i.e. $r(T_1, T_2) = 0$ has been assumed (see Figure 3.8). With reference to the measurement of scattering coefficients, this also suggests that $r(T_2, T_3) = 0$, as the situation is the same as that for T_1 except that during the measurement of T_3 the turntable is rotating. Similarly, it is probable that $r(T_1, T_4) = 0$ and $r(T_3, T_4) = 0$ since the situations are essentially the same as for $r(T_1, T_2)$ and $r(T_2, T_3)$, respectively. These assumptions will be confirmed in Section 3.3.3.C through verification measurements.

With the previous considerations, only $r(T_1, T_3)$ and $r(T_2, T_4)$ remain to be determined. By taking into account the correlation between these input variables the uncertainty of the sample scattering coefficient becomes

$$u_s|_T \approx \frac{2.334}{\sqrt{M\,f_c}} \cdot \frac{K}{1 - \alpha_s} \cdot \sqrt{(1-s)^2 \cdot \sum_{i=1}^{2} \frac{1}{T_i^3} + \sum_{i=3}^{4} \frac{1}{T_i^3} - 2 \cdot A}, \quad (3.67)$$

with

$$A = (1 - s) \cdot \left(\frac{r(T_1, T_3)}{\sqrt{T_1^3 \cdot T_3^3}} + \frac{r(T_2, T_4)}{\sqrt{T_2^3 \cdot T_4^3}} \right). \quad (3.68)$$

Eq. (3.63) can be corrected accordingly. The correlation coefficients will be determined based on the verification measurement described in the next section.

C Experimental Verification

The measurements used to verify the equations that were developed in Section 3.3.3.B were carried out in the small-scale reverberation chamber at the ITA. The dimensions of the rectangular chamber are $1.5\,\mathrm{m} \times 1.2\,\mathrm{m} \times 0.95\,\mathrm{m}$; a photograph is presented in Figure 3.16a. Instead of the typical hanging panels, boundary diffusers have been installed in the chamber, which were created during a study published in [129]. The number of diffusers was determined following the procedure described in Appendix A of ISO 354. Due to the size and shape of the boundary diffusers the room volume and surface area had to be corrected; the corrected values of the configuration used in this study are $V_{\mathrm{room}} = 1.67\,\mathrm{m}^3$ and $S_{\mathrm{room}} = 9.05\,\mathrm{m}^2$.

The chamber is not equipped with an air conditioning system but has a stationary sensor for the measurement of temperature and humidity. The typical relative variation of the climatic conditions during scattering measurements in the small-scale chamber at the ITA is less than $0.5\,\%$ for both temperature and relativity humidity. With regard to the findings concerning the influence of the climatic conditions presented before this leads to the conclusion that in most cases the uncertainty of the climatic conditions does not play a significant role. If larger variations occur they will only have an influence at very high frequencies, which is especially relevant for small-scale environments. However, this will not be pursued further in this thesis.

The turntable needed to carry out scattering measurements has been mounted underneath the floor of the chamber to be able to have the baseplate flush mounted. This is a measure that was taken to avoid an uncertainty caused by the equipment connected to the turning baseplate inside the chamber [143]. As a consequence, the baseplate scattering coefficient is significantly reduced. The baseplate has a diameter of $90\,\mathrm{cm}$ and its edges are covered with a plastic ring (see bottom of Figure 3.16a) to seal the gap between the rotating plate and the rest of the chamber floor.

All measurements described in this section have been performed with a scale factor of $N = 5$ and hence the frequency range was scaled accordingly to cover the third-octave band center frequencies f_c between $500\,\mathrm{Hz}$ and $25\,\mathrm{kHz}$. Nonetheless, to make the data comparable the results presented here have been converted to real-scale, including the frequencies and reverberation times. A total of $M = 10$ independent source-receiver positions have been used.

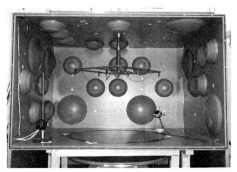

(a) ITA small-scale reverberation chamber

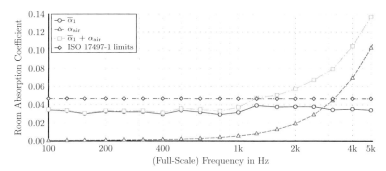

(b) Room Absorption Coefficients of the empty chamber

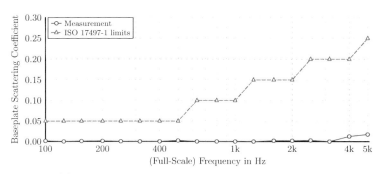

(c) Baseplate Scattering Coefficient of the empty chamber

Figure 3.16: Small-scale reverberation chamber for scattering measurements at the ITA: (a) Setup; (b) Room and air absorption coefficients of the empty chamber with the limits given by ISO 17497-1; (c) Baseplate scattering coefficient together with the ISO limits

In Figure 3.16b the average room absorption coefficient of the empty chamber as a function of frequency is graphed together with the contribution by air attenuation as well as the limits according to ISO 17497-1 (see Eq. (2.107)). The wall absorption is low enough to comply with the conditions set in ISO 174971-1. However the contribution by air attenuation is relatively high which is due to the measurements at the scaled-up frequencies, where the influence of the air attenuation coefficient m is high (see Figure 2.2).

Figure 3.16c shows a typical result of the baseplate scattering coefficient according to Eq. (2.106) together with the ISO limits from Table 2.2 as a function of frequency. It can be confirmed that the measure of moving the turntable out of the chamber is effective in reducing the value of s_{base}, such that it can be neglected in most situations.

The reverberation times (converted to real-scale) from a measurement of a sample with a sinusoidal surface are depicted in 3.17a. The sample, which has been extensively used in other studies [30, 35, 144], has a surface area of $S_{\text{sample}} = 0.55\,\text{m}^2$ leading to a real-scale value of $K = 2.66\,\text{s}$. The fact that $T_3 \approx T_1$ confirms the low values of the baseplate scattering coefficient. The resulting absorption and scattering coefficient of the sample are plotted in Figure 3.17b. The absorption coefficient increases with frequency but remains strictly below 0.4, adhering to the limit of 0.5 demanded by ISO 17497-1. The scattering coefficient starts to rise from 300 Hz and reaches a plateau at 1000 Hz. The theoretically possible maximum value of $s = 1$ is exceeded for frequencies above 3 kHz, which is probably due to uncertainties at high frequencies related to the influence of air attenuation.

To verify the assumptions made before concerning correlation of the reverberation times, the cross-correlation has been evaluated with the corrcoef function in MATLAB. As the correlation coefficient is a symmetric quantity, not all possible combinations had to be tested. The results of the correlation coefficient between T_1 and T_2–T_4 are presented in Figure 3.18a. The correlation of T_2 with T_3–T_4 and of T_3 with T_4 is shown in Figure 3.18b. As for the absorption coefficient before, the significance at the $p = 0.05$ level was also evaluated.

Of the correlation coefficients related to T_1 only $r(T_1, T_3)$ was found to be significantly different from zero, confirming the assumptions mentioned in the last section. Towards higher frequencies, the correlation decreases slightly, which can be attributed to minimal effects of baseplate scattering, however the values remain significant. In order not to complicate a model for the uncertainty and since values of the correlation coefficient above 0.7 are still considered as indicating strong correlation, a value of $r(T_1, T_3) = 1$ will be used in the further analysis.

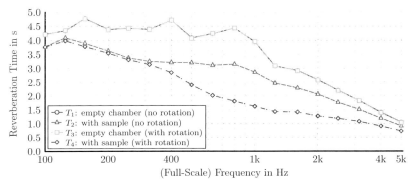

(a) Average Reverberation Time (converted to real-scale)

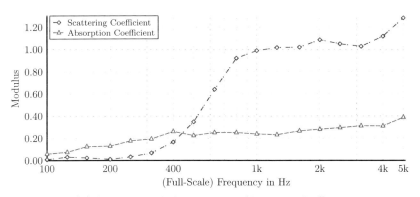

(b) Average Sample Scattering and Absorption Coefficient

Figure 3.17: Average results of (a) the reverberation time and (b) the sample scattering and absorption coefficient for the sinusoid sample used in the verification measurements for the uncertainty of scattering coefficients

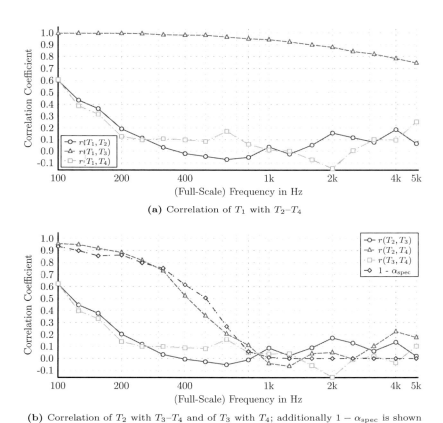

(a) Correlation of T_1 with T_2–T_4

(b) Correlation of T_2 with T_3–T_4 and of T_3 with T_4; additionally $1 - \alpha_{\text{spec}}$ is shown

Figure 3.18: Correlation coefficients between the reverberation times obtained during scattering measurements

Opposed to the results related to the measurement of α_s in Figure 3.8, $r(T_1, T_2)$ and also $r(T_1, T_4)$ increase towards low frequencies. It has already been mentioned before that this could happen for samples with a very low absorption coefficient. Nonetheless, no significant values were found in the frequency range of interest and hence $r(T_1, T_2) = 0$ and $r(T_1, T_4) = 0$ will be assumed.

With the strong correlation between T_1 and T_3 in mind, it does not surprise that the curves for $r(T_2, T_3)$ and $r(T_3, T_4)$ are almost identical to the ones of $r(T_1, T_2)$ and $r(T_1, T_4)$. As the sound field is not significantly different between the measurement of T_1 and that of T_3, it is safe to assume $r(T_2, T_3) = 0$ and $r(T_3, T_4) = 0$.

However there is a strong, frequency-dependent correlation between T_2 and T_4, which is statistically significant and related to the scattering property of the sample. This can be easily explained, considering that for low scattering coefficients the situation corresponds to the case for the empty chamber (T_1 and T_3), where a more or less (acoustically) flat surface is rotated on the turntable. For increasing scattering coefficients, the measurements of the sample in different orientations will correspond to a different sound field and hence the correlation drops.

This can be confirmed by comparing the curve for $r(T_2, T_4)$ in Figure 3.18b with the curve for $1 - \alpha_{\mathrm{spec}}$, which is related to $1 - s$. There is good agreement between the two curves up to 3 kHz, when the measured correlation coefficient rises again, which is probably due to other measurement uncertainties, e.g. air attenuation, that affect all reverberation times equally. Still, for practical purposes, it is safe to use $r(T_2, T_4) = 1 - \alpha_{\mathrm{spec}}$ as there is no statistical significance for the values above 3 kHz.

With the findings regarding the correlation of the reverberation times, the uncertainty according to Eq. (3.67) can be explicitly expressed as

$$u_s|_T \approx \frac{2.334}{\sqrt{M\,f_c}} \cdot \frac{K}{1 - \alpha_s} \cdot \sqrt{(1-s)^2 \cdot \sum_{i=1}^{2} \frac{1}{T_i^3} + \sum_{i=3}^{4} \frac{1}{T_i^3} - 2 \cdot A}\,, \quad (3.69)$$

with (compare Eq. (3.68))

$$A = (1-s) \cdot \left(\frac{1}{\sqrt{T_1^3 \cdot T_3^3}} + \frac{1 - \alpha_{\mathrm{spec}}}{\sqrt{T_2^3 \cdot T_4^3}} \right), \quad (3.70)$$

where the reverberation times T_2–T_4 can be expressed by the room and sample properties as stated before (see Eq. (3.41), Eq. (3.65) and Eq. (3.66)).

The measurement data for the sinusoid sample was finally evaluated concerning the expanded ($C = 1.96$) uncertainty due to the spatial variation of reverberation times. For this, the measurement results obtained at the $M = 10$ source-receiver combinations were evaluated using Eq. (2.113), which gives the actual uncertainty of the sample absorption coefficient. Additionally the evaluation using uncertainty propagation was performed. This was done using the measured uncertainty of the reverberation time according to Eq. (3.63), once without and once with correlation of the input quantities. The approximation following Davy has also been applied taking into account the correlation between the reverberation times (Eq. (3.69)). In the latter case, the modal overlap correction according to Eq. (3.44) and Eq. (3.45) has also been applied. The factors corresponding to u_{T_3} and u_{T_4} are identical with the ones corresponding to u_{T_1} and u_{T_2}, respectively, as the modal sound field at low frequencies is not changed by the rotation of the turning table.

The results of the uncertainty evaluation are presented in Figure 3.19 as a function of frequency. It is evident that by neglecting the correlation coefficients between the reverberation times in the uncertainty propagation (dashed curve with triangle markers), the uncertainty is greatly overestimated compared to the measured values (solid curve with circle markers). This effect is most important for frequencies below 600 Hz, where the scattering coefficient of the sample is low (compare Figure 3.17b). By taking into account correlation (dash-dotted curve with square markers) the prediction of the uncertainty is very close to the measured result. However, at very high frequencies above 4 kHz the predicted values are significantly lower than the measured ones. This is the frequency range where values of the sample scattering coefficient exceed one, indicating influence of an uncertainty in the calculation of air absorption, which has already been mentioned in Section 3.3.3.A.

Compared to the prediction with the measured values of u_T, the approximation following Davy (dashed curve with diamond markers) performs very well, giving very similar values except for the third-octave bands at 1 kHz and 1.25 kHz. The behavior at high frequencies is identical to the one using the measured uncertainty of the reverberation times, which is expected as no additional information on other uncertainty factors has been included.

The correction related to the modal overlap does not have a large impact on the predicted uncertainty in this case (dashed curve with cross markers). The values are practically identical to the ones without the correction. This can be explained by the low absorption coefficient of the sample and the fact that the Schroeder frequency of the chamber $f_s \approx 300$ Hz is relatively low. The maximum values of the correction terms (compare Figure 3.9) are not more than approximately $\frac{u'_T}{u_T} = 1.1$. For samples with a higher absorption coefficient however, the modal overlap correction could become important. This would have to be determined by an extensive study using scattering samples with varying absorption coefficients.

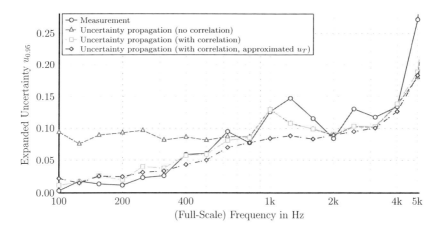

Figure 3.19: Expanded uncertainty of the sample scattering coefficient as a function of frequency evaluated for the sinusoid sample used in this study for a value of $D = 15\,\mathrm{dB}$ and $M = 10$

It can be concluded that the prediction of the uncertainty of the sample scattering coefficient due to a spatial variation of the reverberation times can be performed with the equations developed in 3.3.3.B based on the approximations by Davy. In any case, the correlation between T_1 and T_3 and between T_2 and T_4, respectively, has to be taken into account to avoid an overestimation of the uncertainty.

For the sake of brevity, the validation measurements and evaluation in this section have only been presented for a single sample in one type of reverberation chamber. The good performance of the uncertainty prediction has also been observed for another sample in the same chamber. Nonetheless, further studies have to be carried out in different chambers, especially in full-scale.

D Further Analysis

In analogy to the case of sample absorption coefficients in Section 3.2.3.D, the analysis of the uncertainty of the scattering coefficient will consist of determining a minimum number of source-receiver combinations $M_{\min}$ needed to achieve a given maximum extended uncertainty $u_{0.95,\max}$. The analysis will be performed at $f_c = 100\,\mathrm{Hz}$, which is the lowest frequency considered by the standards, and the desired maximum uncertainty is chosen as $u_{0.95,\max} = 0.1$. From the previous investigations on the uncertainty of scattering coefficient measurements it is obvious that in comparison to the case of α_s the values of $M_{\min}$ will be much larger.

Solving Eq. (3.69) (including the modal overlap correction factors) for $M_{\min}$ gives the following expression:

$$
\begin{aligned}
M_{\min} \geq {}& \left(\frac{1.96 \cdot 2.334}{u_{0.95,\max}} \right)^2 \cdot \frac{K^2}{f_c \cdot (1 - \alpha_s)^2} \\
& \cdot \Bigg[(1 - s)^2 \cdot \left(\left(\frac{u'_{T_1}}{u_{T_1}} \right)^2 \cdot \frac{1}{T_1^3} + \left(\frac{u'_{T_2}}{u_{T_2}} \right)^2 \cdot \frac{1}{T_2^3} \right) \\
& + \left(\frac{u'_{T_1}}{u_{T_1}} \right)^2 \cdot \frac{1}{T_3^3} + \left(\frac{u'_{T_2}}{u_{T_2}} \right)^2 \cdot \frac{1}{T_4^3} \\
& - 2 \cdot (1 - s) \cdot \left(\left(\frac{u'_{T_1}}{u_{T_1}} \right)^2 \cdot \frac{1}{\sqrt{T_1^3 \cdot T_3^3}} + \left(\frac{u'_{T_2}}{u_{T_2}} \right)^2 \cdot \frac{1 - \alpha_{\mathrm{spec}}}{\sqrt{T_2^3 \cdot T_4^3}} \right) \Bigg],
\end{aligned}
$$

$$(3.71)$$

where the expressions relating T_2–T_4 to α_s, s_{base} and s in have not been inserted to keep the expression more compact.

As mentioned in Section 3.3.3.B and as becomes evident from Eq. (3.71) the factors influencing u_s — apart from the frequency f_c and the setup constant K — are the sample absorption coefficient and the scattering coefficient of the baseplate and of the sample. Evaluating the uncertainty for all combinations of these factors would be quite tedious and hence the value of the baseplate scattering will be set to $s_{\mathrm{base}} = 0.025$, which is half of the maximum value allowed by ISO 17497-1 (see Table 2.2).

The value of $M_{\min}$ according to Eq. (3.71) is presented as contour plots in Figure 3.20 and Figure 3.21 as a function of the reverberation time of the empty chamber T_1 and the setup constant K. The graphs in Figure 3.20 and Figure 3.21 have been created for $s = 0.20$ and $s = 0.40$, respectively.

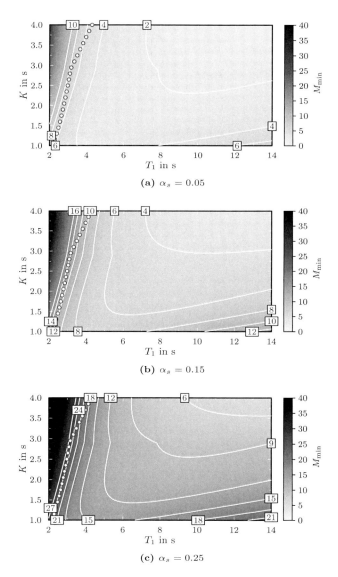

(a) $\alpha_s = 0.05$

(b) $\alpha_s = 0.15$

(c) $\alpha_s = 0.25$

Figure 3.20: Minimum number of source-receiver combinations (according to Eq. (3.71)) needed to achieve $u_{0.95,\mathrm{max}} = 0.1$, calculated at $f_c = 100\,\mathrm{Hz}$ for $s_{\mathrm{base}} = 0.025$, $s = 0.20$ and for different values of α_s

In each case the values of the sample absorption coefficient have been chosen as $\alpha_s = 0.05$, $\alpha_s = 0.15$ and $\alpha_s = 0.25$, representing a range of very low to moderate values at the low frequencies under investigation. In all graphs, the red dotted line represents the minimum values of the reverberation time according to ISO 17497-1.

For a value of the scattering coefficient of $s = 0.20$ at $100\,\mathrm{Hz}$, the ISO recommendation of $M = 12$ seems to be confirmed by the data for $\alpha_s = 0.05$ (Figure 3.20a) and $\alpha_s = 0.15$ (Figure 3.20b). For all combinations of T_1 and K that comply with the ISO limits the minimum number of source-receiver combinations needed to achieve $u_{0.95} = 0.1$ is less than 12.

If the sample is slightly more absorptive, with $\alpha_s = 0.25$ (Figure 3.20c), the values of $M_{\min}$ are roughly doubled, meaning that even for such a relatively low scattering coefficient many more measurement positions might be required. The ISO minimum of $M = 12$ is only sufficient for rooms with $K > 1.5\,\mathrm{s}$ and $T_1 > 5\,\mathrm{s}$.

In comparison, the data evaluated for $s = 0.40$ in Figure 3.21 shows that for a higher value of the scattering coefficient at $100\,\mathrm{Hz}$, the effort to achieve an uncertainty of $u_{0.95} = 0.1$ is significantly higher. For a sample with very low absorption ($\alpha_s = 0.05$, Figure 3.21a) $M = 12$ source-receiver combinations will yield the desired precision in most rooms ($K > 1.25\,\mathrm{s}$ and $T_1 > 4\,\mathrm{s}$), although the reverberation time limits given in ISO 17497-1 already require 50 % more measurement positions. In the case of a more absorptive sample ($\alpha_s = 0.15$, Figure 3.21b), the specified uncertainty can only be achieved in large and reverberant rooms ($K > 3\,\mathrm{s}$ and $T_1 > 6\,\mathrm{s}$) when the reverberation times are measured at $M = 12$ positions. More than twice as many positions ($M_{\min} = 28$) are needed for the minimum reverberation times allowed by ISO 17497-1. For a value of $\alpha_s = 0.25$ (Figure 3.21c), it is not possible to obtain a precise result of the scattering coefficient with $M = 12$ measurement positions in the sound field. In highly damped rooms with short reverberation times that still comply with the ISO limits, at least 40 measurements have to be performed for each of the four reverberation times to achieve $u_{0.95} = 0.1$. This in turn means that the entire measurement procedure would take substantially longer and the uncertainty factors related to the climatic conditions could have a negative effect on the result.

A value of the scattering coefficient of 0.4 at $100\,\mathrm{Hz}$ as used here is relatively high and seldom encountered with real samples. Nonetheless, it has served to establish an impression of a worst-case scenario and to give an example how the formulas derived in this section can be applied in practice.

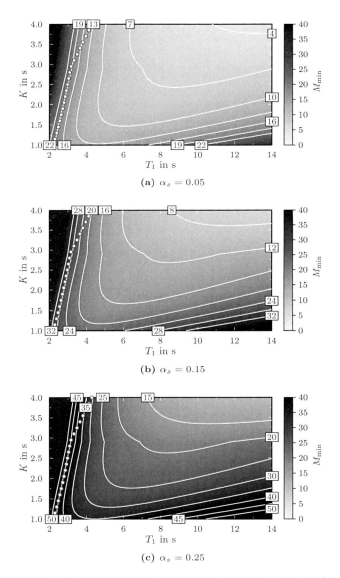

(a) $\alpha_s = 0.05$

(b) $\alpha_s = 0.15$

(c) $\alpha_s = 0.25$

Figure 3.21: Minimum number of source-receiver combinations (according to Eq. (3.71)) needed to achieve $u_{0.95,\text{max}} = 0.1$, calculated at $f_c = 100\,\text{Hz}$ for $s_{\text{base}} = 0.025$, $s = 0.40$ and for different values of α_s

3.4 Preliminary Conclusions

In this section, a thorough investigation was carried out on the standardized measurement methods in the reverberation chamber to measure the sample absorption and scattering coefficient. It was shown how the method of propagation of uncertainty can be used to determine the causes of measurement errors and how they are connected to the properties of the reverberation chamber and the sample.

It was found that an incorrect determination of the room volume and the surface area of the sample leads to a relative error of the sample absorption coefficient that is equal to the relative error of the geometric quantity. For the scattering coefficient the relative error was increased by a factor related to the sample absorption coefficient, effectively increasing the uncertainty. Since it is assumed that the task of determining the volume and surface area is comparatively easy in relation to the measurement of other quantities, the influence was considered to be negligible.

Two systematic deviations that were investigated are related to simplifications in the equations used to calculate the desired coefficients. The use of the Sabine instead of the Eyring equation to determine the room absorption coefficient from the measured reverberation times leads to significant errors especially when measuring the scattering coefficient. The magnitude of this error increases with the room absorption of the empty chamber and with the sample absorption. Another simplification in ISO 354 is that the room surface area covered by the sample is not corrected for. While this is usually not a problem when measuring the sample absorption coefficient, it can become significant for measurements of the scattering coefficient. Both of the errors mentioned due to simplifications in the equations can easily be avoided.

The effect of a variation in the speed of sound was found to be negligible in all practical situations as the climatic conditions would have to vary substantially between measurements in order to produce significant errors. In comparison, the air attenuation coefficient plays a larger role, especially with regard to the measurement of scattering coefficients. As it is not easy to define a simple relationship between m and the climatic conditions, Monte-Carlo simulations have been used to investigate the influence on the room and sample absorption coefficient as well as on the scattering coefficient. The data showed that unless there is an unrealistically large variation of the temperature and/or relative humidity, the effect of an uncertainty of the air attenuation coefficient can be neglected for the sample absorption coefficient. In the case of the scattering coefficient, especially if small-scale measurements are performed, an effect in high frequencies has to be expected so that special care should be taken to ensure stable climatic conditions.

Most emphasis was set on the influence of random variations of reverberation times, especially with regard to spatial fluctuations. For a general analysis, the theoretical predictions of this spatial variation derived by Davy have been applied. The accuracy of the predictions was verified with measurements and very good agreement was found, especially after incorporating the low-frequency correction term related to the statistical modal overlap.

Further analysis showed that the minimum number of source-receiver combinations for the measurement of reverberation times necessary to achieve a given uncertainty in the respective coefficient can be determined from relatively simple equations. In the case of the sample absorption coefficient, the recommendation of ISO 354 to use at least 12 measurement positions leads to acceptable results if the sample is not too absorptive ($\alpha_s \leq 0.5$) at low frequencies.

For the measurement of scattering coefficients a strong relation between the absorption coefficient of the sample and the uncertainty was found. It could be shown that the absorptivity of the sample should be kept to a minimum to obtain reliable results. As for the absorption coefficient, the uncertainty of the scattering coefficient rises with the value of the measured quantity.

A correlation of the reverberation times determined during scattering coefficient measurements was found. The correlation between T_2 and T_4 is related to the scattering properties of the sample. Hence, the equation to predict the uncertainty from the measured reverberation times, which is also derived in Annex A of ISO 17497-1, had to be changed to include correlation. By neglecting this effect the uncertainty is greatly overestimated.

4

Measurement of Angle-Dependent Reflection properties

In Chapter 3 the standardized measurement methods for reflection properties in a diffuse sound field have been analyzed concerning measurement uncertainties. The effort needed to achieve reliable results has been investigated and it was found that it can be hard to precisely measure the absorption and the scattering coefficient of a sample.

Apart from the uncertainties treated in Chapter 3, there are other reasons that may disqualify the use of standardized measurement methods in certain situations. The most obvious is the fact that it may not always be possible to gather a sample with the needed size and dimensions. It may even be impossible to move a sample to the reverberation chamber due to constructive restrictions.

Additionally, a more detailed knowledge of the acoustic behavior of an absorber or scatterer may be desired. Since the standardized methods work with random sound incidence, there is no information on the angle-dependency of the reflected sound. For an appropriate modeling — either of the material or of the reflected sound — it would be beneficial to obtain this information.

With regard to the absorptive properties of a sample, the absorption coefficient may not give enough information for a correct modeling. Instead, the complex reflection factor – or even better, the complex surface impedance — are quantities that should be measured.

In this chapter, the measurement of angle-dependent reflection properties of surfaces will be treated. In this context, both the result of the complex reflection factor as well as the spatial distribution of the reflected sound are of interest. This should lead to results that are useful for exactly modeling sound reflections and in turn the sound field in closed environments.

The equations presented in this chapter will mainly focus on the signal description in the frequency domain. The transform of the data between frequency-domain and time-domain can of course always be performed without a loss of information using the *Fast Fourier Transform* (FFT).

4.1 Measurement Setup

For a simultaneous measurement of the angle-dependent absorbing as well as the scattering properties of a sample, it is necessary to employ multiple sensors at once to obtain an impression on the spatial distribution of the reflected sound. The arrangement of such a sensor array is determined by the possible signal-processing techniques that can be applied to the measured data. The following descriptions have in part already been published in [145]. Part of the design and analysis of the setup has been carried out by ISENBERG [146]. A first numerical uncertainty analysis has been performed by MENDE [147].

In this work, the approach taken is an array of microphones that are distributed on two semicircles with radii of $r_1 = 0.512$ m and $r_2 = 0.527$ m[1] (see Figure 4.1 for a schematic). To reduce the hardware effort, a sequential array is used so that only 24 physical microphones are used to cover the polar angles and a step motor system enables the measurement of many positions in the azimuth direction. The placement of the microphones with regard to the polar angles is carried out in such a way that the complete hemispherical array follows a Gaussian quadrature sampling of order 47 (with regard to the full sphere). This type of sampling is especially suited for a sequential array as it is rotationally symmetric. The advantage of this sampling strategy with respect to the application of the Spherical Harmonics Transform has already been mentioned in Section 2.2.5.

The choice of using a sensor array in the form of two hemispherical shells has several advantages regarding the measurement of angle-dependent reflection properties:

- It is relatively easy to place the center of the array close to a reflecting surface, opposed to complete spheres as in [148].

- Measurements can span all angles of incidence (and reflection), which is not the case for linear or planar arrays.

- The measured data can be used without further processing to visualize the spatial distribution of the sound pressure, which can yield further insight into sound reflection.

- Additionally, the spatial data can be processed in terms of the scattering properties of a sample.

---

[1]The actual radii of the semicircles are 0.565 m and 0.58 m but the microphone mounts lead to the radii of the sensor capsules as stated in the text.

- Spatial filtering methods can be used, e. g. as an alternative to the subtraction method.

- Even if no spatial processing is performed, the large number of measurement positions can be used to return a more stable result for the reflection factor by averaging across receiver positions with the same reflection angle.

The microphones are distributed alternately onto the semicircles so that the two sub-arrays each cover every second polar angle. A schematic of this approach is presented in Figure 4.1, showing the evolution of the microphone positions for step-wise rotation of the array. The SH order of 47 leads to an azimuth resolution of 3.75 degree, which results in 96 turns to achieve a complete rotation yielding a total of 2304 measurement positions on two hemispherical shells.

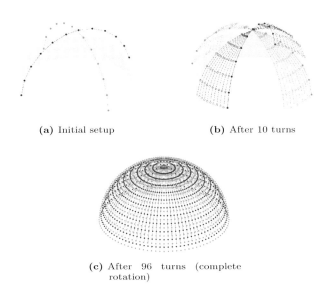

(a) Initial setup   (b) After 10 turns

(c) After 96 turns (complete rotation)

Figure 4.1: Schematic of the use of the sequential array

The reason for choosing different radii for the two semicircles — and thus effectively creating two separate arrays — is to obtain information that can be used in the application of Scattering Near-Field Holography as described

119

in Section 2.2.3. The limiting frequency due to radial aliasing according to Eq. (2.39) for the setup presented here is

$$f_{\text{alias,radial}} = \frac{c}{2 \cdot (0.527\,\text{m} - 0.512\,\text{m})} \approx 11.4\,\text{kHz}\,, \qquad (4.1)$$

which is high considering that usually the uncertainties in the array setup will prohibit the successful application of spherical signal processing for such high frequencies. More importantly, the effect of angular aliasing due to the limited resolution in the polar and azimuth direction reduces the usable frequency range. The discussion of spatial aliasing is not a topic of this thesis and the reader is referred to the work by RAFAELY, WEISS, and BACHMAT [149] and ZOTTER [78] for an overview.

The actual realization of the sequential array with the supporting structure is presented in Figure 4.2 for a measurement setup above a porous absorber in the anechoic chamber of the ITA. The source (grey sphere in the foreground in Figure 4.2) and the microphones used in all measurements are depicted in Figure 4.3a and Figure 4.3b. The respective free-field frequency responses are plotted in Figure 4.3c and Figure 4.3d.

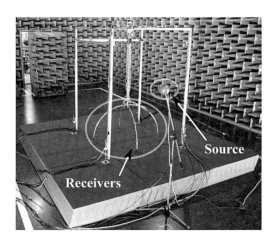

Figure 4.2: Measurement setup in the anechoic chamber of the ITA above a porous absorber

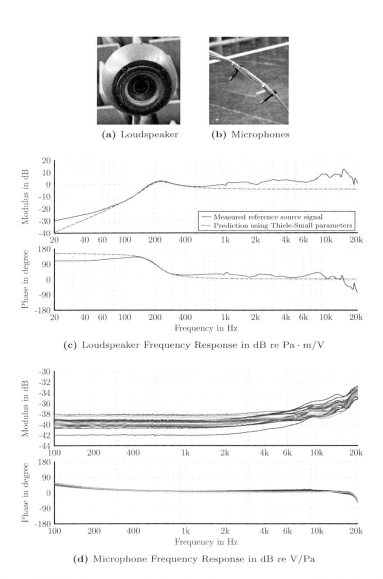

(a) Loudspeaker   (b) Microphones

(c) Loudspeaker Frequency Response in dB re Pa · m/V

(d) Microphone Frequency Response in dB re V/Pa

Figure 4.3: Equipment used for reflection measurements: (a) loudspeaker in spherical enclosure, (b) microphones in array mounts, (c) loudspeaker frequency response together with predicted response using the Thiele-Small parameters, (d) frequency response of all 24 microphones

The loudspeaker consists of a single driver with a diameter of 65 mm in a spherical enclosure. Using a single driver ensures that the acoustical center of the source stays at a stable location and the spherical enclosure results in the least effect due to edge diffraction. This can be confirmed by inspecting the measured frequency response (solid curve in Figure 4.3c), which shows no significant notches throughout the entire audible frequency range.

Since it is hardly possible to obtain an accurate low-frequency result in most environments and time-windowing may also influence the data for low frequencies, the loudspeaker response below the resonance frequency of approximately 200 Hz has been calculated based on the prediction using the Thiele-Small parameters of the loudspeaker [150, 151, 152] (dashed curve in Figure 4.3c). The excellent agreement between measured and predicted levels shows the high accuracy of the reference measurement in the frequency range of a piston-like movement of the loudspeaker membrane. With a calibrated measurement equipment, where all sensitivity coefficients are correctly accounted for, deviations of less than 1 dB are possible.

The frequency responses of the 24 array microphones as depicted in Figure 4.3d show some relative variation in level as well as an increased sensitivity towards high frequencies (above 3 kHz). This suggests that it is imperative to apply an individual wide-band calibration, which is achieved by filtering the actual reflection measurements with the inverted frequency response of each microphone. This has been done for all results presented in this chapter.

4.1.1 Influence of the Measurement Setup on the Sound Field

By inspecting the setup in Figure 4.2 it becomes clear that the sound field captured by the array microphones can be altered by the support structure needed to hold and rotate the sensors. To identify the influence and determine possible improvements of the setup a measurement series was conducted in the anechoic chamber.

The influence of the array and its support structure was investigated by measuring the transfer function between the source and eight distributed receivers placed on the reflective floor of the chamber underneath the array under different conditions. All measurements were referenced to the case where the chamber was empty except for source and receivers. The following scenarios have been measured (photographs are presented in Figure 4.4):

1. Support structure only

2. Support structure with microphone array (as in Figure 4.2 without the absorber)

3. Support structure only, foam attached to the entire support structure (Figure 4.4a)

4. Support structure with microphone array, foam attached to the entire support structure (Figure 4.4b)

5. Support structure only, foam attached only to the top parts of the support structure (Figure 4.4c)

6. Support structure with microphone array, foam attached only to the top parts of the support structure (Figure 4.4d)

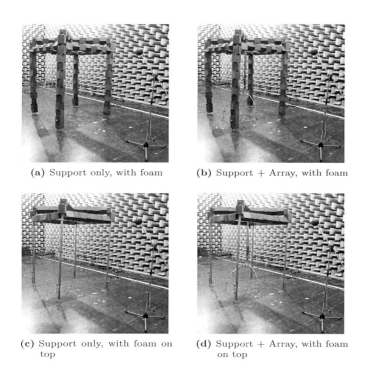

(a) Support only, with foam   (b) Support + Array, with foam

(c) Support only, with foam on   (d) Support + Array, with foam
top   on top

Figure 4.4: Measurement scenarios to determine the influence of the array support structure on the sound field

The foam was attached to the structure of the array to reduce the influence at high frequencies, where the wavelength has a similar dimension as the pieces of the support structure.

The measurements were evaluated by dividing the measured transfer functions to the reference case and then taking the maximum deviation in dB across the eight receivers. In the ideal case of no influence the curves should lie at 0 dB. The results for the six scenarios mentioned above are shown in Figure 4.5 as a function of frequency.

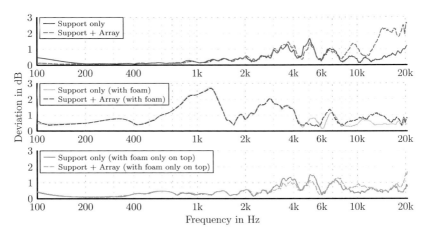

Figure 4.5: Influence of the array support structure on the sound field

The data for the setup without any foam (top graph in Figure 4.5) shows that the influence of the equipment on the sound field is less than 1 dB throughout most of the frequency range. As expected, adding the array to the setup (dashed curve) increases the influence in the high frequencies. In the medium frequency range $(3 - 8 \text{kHz})$ the support structure seems to have the biggest influence, leading to deviations to the free-field result of up to 1.5 dB.

Contrary to the desired effect, attaching foam to the support structure actually increases the deviations between 400 Hz and 5 kHz (middle graph in Figure 4.5). This can be explained by the fact that at low and medium frequencies, where the absorptivity of the foam is relatively low, the added material effectively increases the dimensions of the structure so that the influence becomes larger and is shifted to lower frequencies.

Nonetheless, at high-frequencies (above 10 kHz) the foam has the desired positive effect of attenuating the sound reflected and diffracted by the support structure and the array, which leads to deviations of less than 1 dB.

By only covering the top part of the support structure (bottom graph in Figure 4.5) the desired result is achieved that the high-frequency deviations are reduced by the absorptive properties of the foam but the deviations in the low and medium frequency are not increased. This configuration has then be used for all measurements.

4.1.2 Determining Source and Receiver Positions

Especially for the application of spatial signal processing, an exact knowledge of the receiver positions is crucial in obtaining sensible results. The effect of misplaced sensors concerning beamforming calculations has already been investigated in [153]. It was found that above approximately 6 kHz an average misplacement of the microphones of more than 10 mm can have a significant effect.

To obtain accurate results also for high frequencies for the measurements presented in this thesis, the microphone (and source) locations have hence been optimized based on the acoustic propagation times τ, determined from the starting times of the measured impulse responses (see also [154]). These were compared to the predicted propagation times based on the assumed Cartesian position vectors of the loudspeaker $\mathbf{L}$ and microphones $\mathbf{M}$ to determine the distance error Δd for each microphone[2]:

$$\Delta d_i = \|\mathbf{L} - \mathbf{M_i}\| - \tau_i \cdot c \,, \tag{4.2}$$

using the speed of sound c. Here, $\|\ldots\|$ denotes the vector norm.

An optimization based on Eq. (4.2) consists of determining the translation vectors $\mathbf{\Delta L}$ and $\mathbf{\Delta M_i}$ for the loudspeaker and the microphone with index i, respectively that minimize the distance error[3]:

$$\operatorname*{arg\,min}_{\mathbf{\Delta L},\mathbf{\Delta M_i} \in \mathbb{R}^3} \|(\mathbf{L} + \mathbf{\Delta L}) - (\mathbf{M_i} + \mathbf{\Delta M_i})\| - \tau_i \cdot c \,. \tag{4.3}$$

The direct application of Eq. (4.3) to the set of 2304 measured responses cannot yield a correct result as the system is under-determined: the number of degrees of freedom for the optimization of the 3-element translation vectors of the loudspeaker and microphones is $N_{\mathrm{dof}} = 3 + 2304 \cdot 3 = 6915$, whereas only 2304 values are known for τ. However, due to the setup of the array the translational vectors for the microphones are not independent.

---

[2] This is based on the assumption that the starting times of the impulse response correctly represent the actual propagation times. As this has given good results for the measurement of the reference source signal, the assumption seems justified.

[3] It is assumed here that the speed of sound is known through measurements of temperature and humidity. Any uncertainty in the determination of c of course influences the accuracy of the optimization approach

In fact only 24 unique translation vectors exist for the microphones as the rotation of the physical sensors leads to the final positions. A model of the sensor movement has thus been established that takes into account that the support structure may be translated or tilted and that the initial setup of the array may be incorrect. Using this model, the effective number of degrees of freedom for the optimization can be reduced to $N_{\mathrm{dof}} = 80$. The system to be optimized is then over-determined and a stable solution can always be found.

An exemplary result of the distance error in millimeter is depicted in Figure 4.6 for the initial error (dashed curve) and the result after a least-squares optimization process (solid curve). The good performance of the optimization using the model for the microphone locations can be seen from the fact that the error can be kept below 5 mm for almost all receiver positions. This is a sub-sample accuracy as one sample at 44.1 kHz sampling rate corresponds to a distance of 7.8 mm. The average absolute distance error of approximately 10 mm for the initial positions shows that the loudspeaker position was not estimated correctly. After the optimization, the average absolute error is reduced to 1 mm.

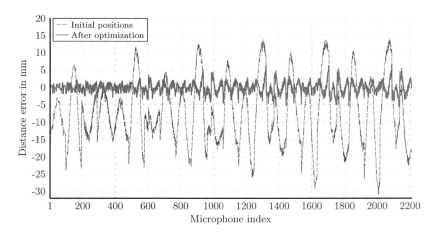

Figure 4.6: Distance error according to Eq. (4.2) in mm before and after optimization

It has to be mentioned of course that any errors in the setup that are not included in the optimization model can also not be corrected. The good results, however, do not suggest that there is another large source of error that has been neglected.

4.1.3 Measurements with a Calibrated Source

It will be assumed from this point on, that measurements are performed with a calibrated source. This means that the free-field transfer function $\underline{H}_p$ of the sound pressure at a certain distance r_0 from the loudspeaker relative to the input voltage at the loudspeaker terminals has been accurately measured. The calibrated source transfer function $\underline{H}_s$ is then obtained by compensating for the propagation between source and receiver:

$$\underline{H}_s(k) = \frac{\underline{H}_p(k, r_0)}{\underline{G}(k, r_0)} \,, \tag{4.4}$$

where $\underline{G}(r)$ is the Green's function for free-field propagation (see Eq. (2.18)). The unit of $\underline{H}_s$ is $\mathrm{Pa \cdot m/V}$ and for purely spherical radiation $\underline{H}_s(k) = \mathrm{j}\omega\rho_0\underline{Q}(k)$ (compare Eq. (2.17)). An example for the source used in this study has already been presented in Figure 4.3c.

With the calibrated source transfer function, the resulting incident sound pressure at any distance r from the source can be calculated — neglecting the directivity or any near-field effects — by

$$\underline{H}_{\mathrm{inc}}(k, r) = \underline{H}_s(k) \cdot \underline{G}(k, r) \,, \tag{4.5}$$

The approach of using a calibrated source has the advantage that the incident sound at the microphone can be predicted based only on the knowledge of the distance between the source and the microphone. This information can be determined quite accurately from the acoustic impulse response (see Figure 4.6), under the assumption that the hardware latency has been calibrated and accounted for.

There are, however, also drawbacks to this approach. Obviously, if the source should behave differently between the calibration measurement and the actual reflection measurement — e. g. non-linearities caused by excessive input voltages — an error would be made by applying Eq. (4.5). A similar influence can be related to a change of the climatic conditions influencing the speed of sound. Most importantly, any effects related to the near-field radiation and the directivity of the loudspeaker are completely disregarded in this approach.

The influence of near-field effects can be reduced by demanding that the reference measurement in the free-field should always be performed at approximately the distance that is intended for the reflection measurements. This, however, still leaves the issue of the source directivity. A possible solution to this problem is presented in Section 4.1.4.

4.1.4 Including Source (and Receiver) Directivity

As was mentioned in the last section, a possible source of error when measuring with a calibrated source is the directivity of the loudspeaker, which would be neglected if a point-like wave propagation is assumed. This actually does not only hold for the source but also for the microphones as receivers. Whether the influence of the directivity is large and how it can be compensated for will be investigated in this section. The method described here — from now referred to as *monopole decomposition* — has already been presented in [155] and it has been investigated in detail by BRAREN [156].

To justify the need to incorporate a model for the source directivity, the isobars of the measured directivity of the loudspeaker shown in Figure 4.3a are presented in Figure 4.7 as a function of frequency. The directivity has been measured in the anechoic chamber of the ITA on a spherical grid of radius 2 m following a Gaussian sampling scheme of order $N_{\mathrm{max}} = 40$, resulting in $M = 3362$ measurement positions. The data has been normalized with respect to the frontal (*on-axis*) direction, i. e. to the response plotted in Figure 4.3c. The results are shown as a function of the elevation angle (cut through the xz-plane) in Figure 4.7a and as a function of the azimuth angle (cut through the xy-plane) in Figure 4.7b. The contour lines are drawn at levels of $-3\,\mathrm{dB}$ and $-6\,\mathrm{dB}$ and then in steps of $10\,\mathrm{dB}$ from $-10\,\mathrm{dB}$ to $-60\,\mathrm{dB}$.

It can clearly be seen from the data that the half-power beam-width (difference between the contour lines at $-3\,\mathrm{dB}$) decreases monotonously up till $10\,\mathrm{kHz}$, where the width is approximately 30 degree. Above $10\,\mathrm{kHz}$ the directivity varies significantly, especially at $14\,\mathrm{kHz}$, probably due to a membrane mode. This shows that the loudspeaker does not radiate purely spherical waves like a point source at high frequencies. As the angle between the frontal direction and the receiver becomes larger, the influence of the directivity will become significant at lower frequencies.

For the application in the measurement scenario depicted in Figure 4.2, the schematic in Figure 2.9b regarding the reflection of spherical waves can be modified to include directive sources. This will make clear how the measurement of reflection factors is affected by the directivity. In Figure 4.8 the modified schematic is shown for the directive source and a single receiver. The source is oriented with its main axis toward the center of the coordinate system.

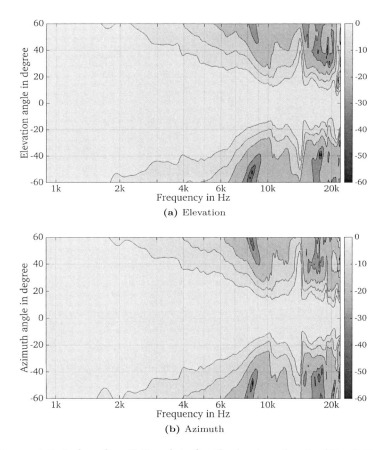

(a) Elevation

(b) Azimuth

Figure 4.7: Isobar directivity plots for the loudspeaker in dB relative to frontal direction; the first three contour lines are drawn at levels of $-3\,\mathrm{dB}$, $-6\,\mathrm{dB}$ and $-10\,\mathrm{dB}$

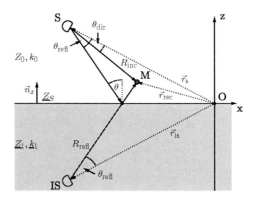

Figure 4.8: Modified schematic for sound reflection of spherical waves (compare Figure 2.9b) including a real loudspeaker

Two important conclusions can be drawn from the schematic in Figure 4.8: (1) opposed to the situation with point sources, the orientation of the source has a direct influence on the received signal, and (2) the angle measured from the frontal direction of the loudspeaker is different for the direct sound received by the microphone (θ_{dir}) and the reflected sound radiated by the image source (θ_{refl}). The latter observation is important for the determination of reflection factors because it means that the directivity of the source does not cancel out.

Assuming a symmetric radiation pattern, the loudspeaker could of course be oriented in such a way that the angles for the direct and reflected sound are approximately equal, i.e. $\theta_{\mathrm{dir}} \approx \theta_{\mathrm{refl}}$. However, in practice this is not easy to achieve and it is especially impossible when employing multiple receivers at the same time. This is why the loudspeaker used in the measurements with the hemispherical array has always been oriented towards the center of the array as a fixed reference position.

In [155] it was shown that the source directivity can have a significant impact on the result of reflection factors if the angle between the loudspeaker axis and the receiver is large. Following the approach by ESCOLANO, LÓPEZ, and PUEO [157], the loudspeaker was modeled as a cluster of monopole sources with frequency-dependent weighting filters. In the application of sound reflection, this is beneficial as the mathematical reflection models are based on point sources so that the sound field including the directivity of the loudspeaker can be calculated as the superposition of the contribution of the substitute sources.

To determine the monopole decomposition of the loudspeaker (or microphone), the following steps have to be performed:

1. Measurement of the directivity (i. e. the sound pressure vector $\mathbf{p}$) of the real source at M positions

2. Specification of the positions of N substitute point sources, with $N \ll M$

3. Determination of the frequency-dependent vector of weights $\mathbf{q}$ for the substitute sources

The weighting filters for the substitute sources can be determined based on the following considerations: the sound pressure $\mathbf{p}$ is related to the weights $\mathbf{q}$ of the substitute point sources through the Green's function (Eq. (2.18)). Arranging all combinations of the transfer function between the substitute sources and the measurement positions for the directivity into a matrix $\mathbf{G}$ containing the Green's function terms

$$\mathbf{G} = \begin{pmatrix} \frac{1}{4\pi r_{1,1}} e^{-jkr_{1,1}} & \cdots & \frac{1}{4\pi r_{N,1}} e^{-jkr_{N,1}} \\ \vdots & \ddots & \vdots \\ \frac{1}{4\pi r_{1,M}} e^{-jkr_{1,M}} & \cdots & \frac{1}{4\pi r_{N,M}} e^{-jkr_{N,M}} \end{pmatrix} \tag{4.6}$$

gives the simple relationship

$$\mathbf{p} = \mathbf{G}\,\mathbf{q}\,. \tag{4.7}$$

In Eq. (4.6) $r_{i,j}$ is the distance between substitute source i and the directivity measurement position j. With the help of Eq. (4.7), the source weights can be calculated with the Tikhonov-regularized inverse as

$$\hat{\mathbf{q}} = \left(\mathbf{G}^H \mathbf{G} + \lambda \mathbf{I}\right)^{-1} \mathbf{G}^H \mathbf{p}\,. \tag{4.8}$$

For the loudspeaker used in the setup described here, the substitute sources were distributed along the coordinate axes with a maximum radius of $r_{\max} = 4\,\mathrm{cm}$. The regularization parameter was chosen as $\lambda = 10^{-6}$ independent of frequency. The average relative error in percent across all measurement positions between the original directivity and the synthesis with monopole decomposition is presented in Figure 4.9 as a function of frequency for three different numbers of monopoles. Additionally, the error limit of 12 % — corresponding to a level change of 1 dB — is graphed.

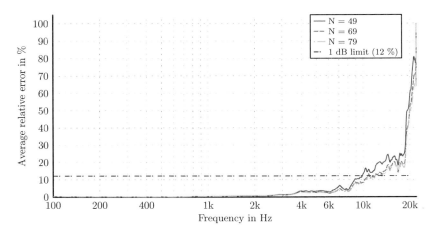

Figure 4.9: Average relative error in percent of the monopole decomposition
according to Eq. (4.8) with $\lambda = 10^{-6}$ and $r_{\mathrm{max}} = 4\,\mathrm{cm}$

A very good performance of the monopole decomposition can be achieved for
all three numbers of substitute sources up to 8 kHz. Above this frequency
the directivity of the loudspeaker becomes more complex. The error can
be reduced by increasing N from 49 to 69, effectively shifting the limiting
frequency from 10 kHz to approximately 14 kHz. However, a further increase
from $N = 69$ to $N = 79$ does not yield a much better accuracy. Hence, for the
use in determining the reflection factor, a value of $N = 69$ has been chosen.

Certainly, a further optimization of the location of the substitute sources could
be carried out to achieve an even better result at high frequencies. However,
with regard to the practical application, the current upper frequency of 14 kHZ
is already considered high. It should also be mentioned that the performance
of the monopole decomposition depends on an accurate measurement of the
original directivity. The numerical accuracy during the synthesis, especially
regarding the location of the substitute point sources, is also important as
the method is very sensitive to phase errors.

A completely different approach to include the directivity into the model could
be pursued through a description in the SH domain. By employing radial
filters, near-field effects could then also be accounted for. However, in the
practical application this would be much more complex as the translation and
rotation operators in the SH domain, which are not easy to calculate, would
have to be used to move the source to the actual position in the reflection
measurement setup (see [76, Chapter 3] and [158]). This is much easier

performed with linear algebra manipulations of the positions of the substitute point sources.

Although the method of monopole decomposition has been applied here to the loudspeaker as an example, the same procedure and mathematical processing steps can be performed to model the directivity of the microphones. Since usually microphones — especially the small capsules used here with a membrane diameter of approximately 4 mm — are not very directive, the more important application definitely concerns the loudspeaker as the sound source. This can be confirmed by looking at the isobar plot for one of the microphones presented in Figure 4.10. As the microphone housing is rotationally symmetric, only the data along the elevation angle is presented. The contour lines are drawn in steps of 1 dB from −1 dB to −6 dB and the data has been normalized to the frontal direction.

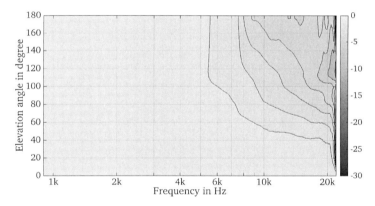

Figure 4.10: Isobar directivity plot for the microphone in dB relative to frontal direction; the first three contour lines are drawn at levels of −1 dB, −2 dB and −3 dB

The isobar plots show that in the primary frequency range of interest up to 5 kHz the microphone can be considered as a perfect monopole. For higher frequencies the level decreases only moderately for increasing elevation angles. This means that in most situations an effect due to the directivity of the microphones is not to be expected. However, because in the setup used in this study the angle of sound incidence towards the microphones can be as large as 180 degree the microphone directivity will also be taken into account using monopole decomposition.

4.2 Separation of Incident and Reflected Sound

The first step in determining reflection properties from measurements consists of obtaining the incident sound and the sound that is reflected from the surface separately. With the appropriate models of sound reflection this information can then be converted into the desired reflection properties (see Section 4.3 and Section 4.4).

With respect to the application in reflection measurements, there are several solutions to the problem of separating the incident and reflected sound pressure. These especially depend on whether a single microphone is employed or the measurement setup involves an array of sensors, which also determines the domain the signal-processing is performed in:

- **Single Microphone**: When using a single microphone or few distributed microphones, the only information available is that of the temporal (and spectral) characteristics of the signal and of the propagation delay, which in turn can be used to determine the distance between source and receiver. The separation can then be performed in one of two ways. The choice depends on whether the reference source signal is available or not:

  Subtraction Method (Time-Domain): If the reference signal can be used to predict the incident sound at the microphone, this information can be subtracted from the recorded pressure to obtain an estimate of the reflected sound. The subtraction method is probably the most commonly used approach to separate signals obtained during reflection measurements [53, 60, 159] (see also the literature review by GEETERE [52]). A detailed description of this method will be given in Section 4.2.1 and Section 4.2.2.

  Windowing (Time-Domain): If no information at all is available on the source signal, the incident and reflected sound have to be separated by time-windowing. This approach has many disadvantages, especially for receivers close to the reflecting surface. In that case the impulses of the incident and reflected sound will arrive at the receiver with a short time difference. Thus, very short time-windows would be needed leading to a loss of low-frequency information [125]. This will be further discussed in Section 4.2.3.

- **Array of Microphones (Spatial Domain)**: For measurements performed with a sensor array, spatial filtering techniques can be applied for a separation of the incident and reflected sound. By moving the signal processing into the spatial domain, especially the low-frequency loss caused by time-windowing is avoided. The data processing depends on the spatial distribution of the sensors. This will be described in more detail in Section 4.2.4.

4.2.1 Subtraction Method (Time-Domain)

The obvious solution to removing the incident sound from a reflection measurement is to subtract it (compare to Eq. (2.67) and Eq. (2.72)). With a source reference measurement and the knowledge of the positions of the source and the receiver the sound pressure is completely determined.

Under ideal circumstances, the subtraction process should leave only the reflected sound. An example is presented in Figure 4.11 for a loudspeaker (modeled as a point source) at a height of 0.75 m above a locally reacting porous absorber with an impedance as depicted in Figure 2.7. The height of the receiver is 0.25 m and the angle of incidence is at 45 degrees. Additionally to the superposition of incident and reflected sound (solid curve), the negative incident sound (dashed curve) and the result of the ideal subtraction process (dash-dotted curve) are shown.

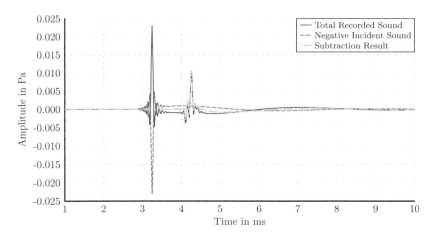

Figure 4.11: Ideal result of the subtraction method, simulated for an angle of incidence of 45 degrees for a source 0.75 m above a locally reacting surface with an impedance as depicted in Figure 2.7; the receiver is 0.25 m above the surface

It can be seen that in the ideal case the incident sound is completely removed and only the reflected sound remains. Obviously, such a perfect result is rarely achieved. Steps that can be taken in that case will be described in the next section.

4.2.2 Optimized Subtraction Method (Frequency-Domain)

The optimal subtraction result can in practice rarely be achieved. This can have many causes that are mostly related to a change of the measurement environment between the determination of the reference and the actual reflection measurement. To give an example, a change of the climatic conditions changes the speed of sound and if this is not correctly accounted for the propagation times will be calculated incorrectly. Other sources of error include an inaccurate determination of the source and receiver locations or calibration errors. Both of these types of error can result in amplitude and phase deviations of the reference signal in relation to the recorded signal above the reflecting surface.

To improve the subtraction result in such cases, an optimization approach based on the work by ROBINSON and XIANG [160] has been implemented. The optimization variables are a constant gain factor and a sub-sample time-shift, which lead to the optimal subtraction result. If many microphones are used in the measurement, the optimization as described in [160] is very time-consuming, as many oversampling rates have to be tested. This is why here the sub-sample shifts are performed in the frequency domain.

With the target function as defined in [160] the optimization would again be relatively slow, as many transforms between the time- and frequency-domain would have to be performed. The approach implemented here is different in so far as not only a part of the time-domain signal is considered for the target function. Instead, the idea is that for a correct subtraction the energy of the entire signal must be minimal. Using the connection between the energy of the time-domain signal $s(t)$ and the frequency-domain equivalent $S(f)$ defined by Parseval's theorem [88, Section 2.9]

$$\int_{-\infty}^{\infty} |s(t)|^2 \, \mathrm{d}t = \int_{-\infty}^{\infty} |S(f)|^2 \, \mathrm{d}f \, , \tag{4.9}$$

the optimization goal for the subtraction result $\underline{H}_{\text{tot}}(f) - \underline{H}_{\text{inc}}(f)$ can be formulated in the frequency-domain as

$$\underset{\Delta t \in \mathbb{R}, A \in \mathbb{R}_{>0}}{\arg\min} \int\limits_{-\infty}^{\infty} \left| \underline{H}_{\text{tot}}(f) - A \cdot \underline{H}_{\text{inc}}(f) \cdot e^{-j\,2\,\pi\,f\cdot\Delta t} \right|^2 \, \mathrm{d}f. \qquad (4.10)$$

Here, $\underline{H}_{\text{tot}}(f)$ is the transfer function of the superposition of incident and reflected sound and $\underline{H}_{\text{inc}}(f)$ is the predicted transfer function of the incident sound. The result of the optimization will be the time-shift Δt and the constant gain A that lead to a minimum residual energy in the subtraction. The idea of optimizing the reference is certainly not intended to compensate for a severely inaccurate calibration or determination of the source and receiver positions. This is why the optimization variables are bounded to $\frac{1}{\sqrt{2}} \leq A \leq \sqrt{2}$ and $-\frac{2}{44100} \leq \Delta t \leq \frac{2}{44100}$.

To obtain an impression of the influence of inaccuracies on the subtraction result as well as the performance of the optimization process, the example shown in Figure 4.11 has been used but the level of the reference signal was changed by $-1\,\mathrm{dB}$ to simulate a calibration error. Additionally, the receiver location used to calculate the reference signal was changed by $4\,\mathrm{mm}$. The result of the subtraction process before and after optimization is shown in Figure 4.12a and Figure 4.12b, respectively. In the figures, the predicted incident sound has been shifted by $-1\,\mathrm{ms}$ to better show the remaining residual signal.

The data in Figure 4.12a clearly shows that the subtraction process was not successful as a substantial part of the signal remains in the result. Especially in relation to the reflected impulse (after $4\,\mathrm{ms}$), the residual incident signal is only approximately $6\,\mathrm{dB}$ lower in amplitude. This will definitely cause a comb-filter effect in the frequency-domain.

In comparison, the optimization according to Eq. (4.10) yields a much better subtraction result (see Figure 4.12b). The incident signal has been attenuated by approximately $20\,\mathrm{dB}$ relative to the result without optimization. On close inspection it can be seen that some residual signal still remains after the optimization and this will generally be the case as the optimization cannot ensure a perfect subtraction.

The remaining signal has mostly high-frequency content as was also observed by MOMMERTZ [53]. This high-frequency content can be removed with a left-sided time-window that only leaves the reflected impulse. In most practical situations this procedure of optimization and subsequent time-windowing will give a good result.

137

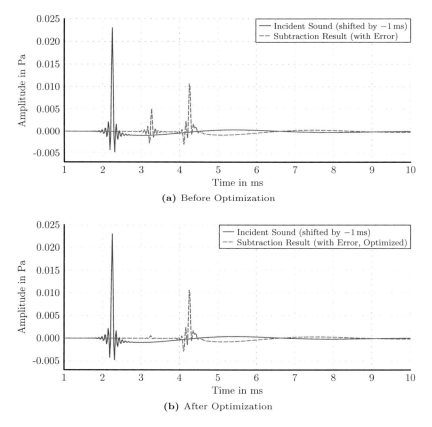

(a) Before Optimization

(b) After Optimization

Figure 4.12: Result of the subtraction method for an amplitude error of the reference signal of $-1\,\mathrm{dB}$ and a misplacement of the receiver of $4\,\mathrm{mm}$; the setup is the same as in Figure 4.11

4.2.3 Windowing (Time-Domain)

In some cases it may not be possible to separately obtain a reference signal for the source. A possible solution is then to determine an *in-situ* reference by separating the impulses in the time-domain with a window function. An application to the example used before is depicted in Figure 4.13.

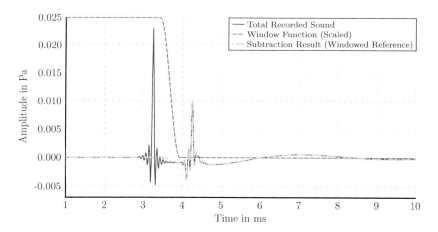

Figure 4.13: Result of the subtraction method when using a windowed (*in-situ*) reference; the setup is the same as in Figure 4.11

The window function has to be automatically adjusted based on the positions of source and receiver. This of course means that the window will be very short for situations where the receiver is close to the reflecting surface (in the example, the difference in the times of arrival is 0.86 ms). Apart from the obvious loss of low-frequency content for the reference signal this can lead to problems if the window-time is significantly shorter than the length of the loudspeaker impulse response. In that case, there is an overlap of the impulses from the incident and reflected sound in the time-domain, which causes low-frequency signal content of the incident sound to remain in the subtraction result.

The overlap of the source signal with the reflected signal can be confirmed with the data in Figure 4.13. The result of the subtraction is obviously affected by the low-frequency decay of the incident sound. This becomes much clearer when looking at the data in the frequency-domain.

In Figure 4.14 the results of all the subtraction methods described before are plotted in the frequency-domain. For the sake of clarity the curves have been shifted by -10 dB relative to each other.

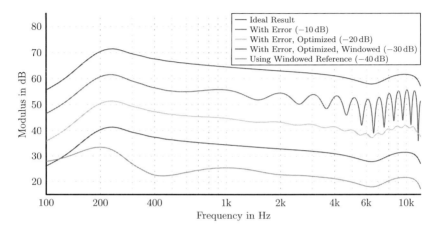

Figure 4.14: Result of the subtraction method in the frequency-domain for different processing methods; the setup is the same as in Figure 4.11

Compared to the ideal result (top curve, compare Figure 4.11) the approach of using the windowed reference (bottom curve) shows large deviations in the low and medium frequency range. A good match with the correct result can only be achieved above approximately 2 kHz.

The effect of the optimization process described in Section 4.2.2 can be clearly seen in the frequency-domain. Without the optimization (second curve from top) a high-frequency comb filter can be observed in the subtraction result. After the optimization (third curve from top) the comb filter is significantly reduced but still present. After the subsequent time-windowing (second curve from bottom) there is an almost perfect match to the ideal result.

It can thus be concluded that working with a calibrated source is the preferred method when using the subtraction method. The presented optimization can to some extent improve the results under practical conditions when the setup is not exactly known. This will be further validated with real measurements in Section 4.3. An *in-situ* reference through time-windowing can only be obtained if the difference of the time of arrival between the incident and reflected sound is large enough to provide correct information in the medium and low frequency range.

Actually, a combination of the presented approaches yields the best results as the influence of the directivity and slight sensor misplacement might deteriorate results with a calibrated source for high frequencies. This has been shown by DIERKES [161] who also investigated several other uncertainty factors influencing the subtraction result. Thus using the *in-situ* (i. e. windowed) reference for high frequencies can be a viable alternative.

4.2.4 Spatial Filtering (SH-Domain)

If the spatial distribution of the sound pressure has been measured with an array of sensors, spatial filtering techniques such as holography or beamforming can be applied. Especially concerning the low-frequency performance this can be advantageous compared to the time-windowing method.

It has been shown in [162] that the method of Scattering Near-Field Holography could be used to separate the waves impinging onto an array from the ones coming from inside the array. However — as was also mentioned then — the method is only defined for data on complete spherical shells (see Section 2.2.3). In this section, it will be analyzed whether the method for base functions on incomplete spheres in Section 2.2.5 can be applied to this problem to enable array-processing in the SH domain on hemispheres, from here on called the *Hemi-Spherical Harmonics* (HSH) domain. Some of the following results have been presented in [163].

It should be mentioned at this points that for the application of measurements of the reflection factor of large surfaces holographic methods are only partially suited. This can be easily seen by inspecting the image source method (Eq. (2.72)) for modeling the reflection from surfaces. The contribution by the image source is also coming from outside the array. Hence, holographic methods can in that case only yield the pressure due to the incident pressure together with the contribution by the image source. Nonetheless, spurious reflections that may occur within the array can still be removed with this technique. Further processing with beamforming or plane-wave-decomposition methods can of course be applied to the result of the holography method to spatially separate the incident and reflected sound pressure.

A situation that is more suited for the application of holographic methods is the measurement of samples that are smaller than the array dimensions. This is usually the case when the scattering properties are of interest. Especially with regard to the measurement of diffusion coefficients, which usually involves the subtraction method, the advantages of holographic methods can be exploited. This will be explored in Section 4.4.

In the following analyses and examples, the two sub-arrays of the Gaussian sampling scheme described in Section 4.1 (see Figure 4.1) are used. The maximum order for the SH transform is chosen as $N_{\mathrm{max}} = 23$, which results in $N_{SH} = (N_{\mathrm{max}} + 1)^2 = 576$ base functions. To determine a frequency limit for spatial aliasing, the so-called kr-limit can be employed [149, 158]. It was shown in [164] that at least two additional orders are necessary to achieve an acceptable sound-field description. Hence, an approximation for the angular aliasing frequency can be given by

$$f_{\mathrm{alias,angular}}(r) \approx \frac{c}{2\,\pi} \cdot \frac{N_{\mathrm{max}} - 2}{r} \,, \qquad (4.11)$$

which results in a value of $f_{\mathrm{alias,angular}} \approx 2.2\,\mathrm{kHz}$ for the arrays used in the examples. The angular aliasing frequency for the combined array is $f_{\mathrm{alias,angular}} \approx 4.68\,\mathrm{kHz}$. As mentioned before, this is significantly lower than the radial aliasing frequency of $11.4\,\mathrm{kHz}$ according to Eq. (4.1).

A Orthonormal Base Functions for the Hemisphere

In a first step, the orthonormal base functions on the hemisphere $\hat{\underline{Y}}_{n'}^{m'}$ — i. e. the Hemispherical Harmonics — will be determined and it will be analyzed how these base functions and the coefficients determined with them can be related to the original base functions, i. e. the Spherical Harmonics.

The procedure described in Section 2.2.5 has been carried out for the spatial sampling mentioned before. The truncation parameter for the SVD was $2 \cdot 10^{-3}$ reducing the number of base functions to $\hat{N}_{HSH} = \left(\hat{N}_{\mathrm{max}} + 1\right)^2 = 324$ for a maximum order of $\hat{N}_{\mathrm{max}} = 17$. The truncation parameter was chosen to obtain the least mixing of the Spherical Harmonics in the new base functions. The method will be described later in this section. The inverse of the matrix $\mathbf{R}$ to establish the relationship between the bases (see Eq. (2.52)) has been determined by the pseudo-inverse based on singular value decomposition, from here on denoted by $\mathbf{R}^{+}$.

To obtain an impression of the shape of the hemispherical base functions, the first 16 complex functions calculated according to Eq. (2.52) are shown in Figure 4.15. In these plots, the radius and color correspond to the modulus and phase, respectively. It seems that the organization into orders and degrees is possible also for the hemispherical base functions and hence this organization will be used in all following descriptions.

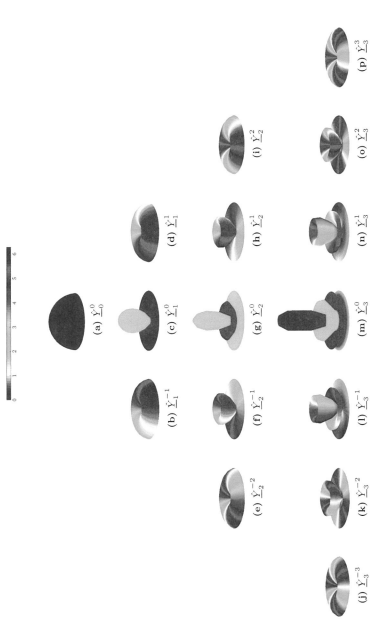

Figure 4.15: First 16 complex-valued Hemispherical Harmonics; the radius and color correspond to the modulus and phase, respectively

The analysis of the relationship between the original base functions and the ones calculated on the bounded domain is based on the matrix $\mathbf{R}^+$ with a size of $\left[(N_{\max} + 1)^2 \times \left(\hat{N}_{\max} + 1 \right)^2 \right]$. Each column of $\mathbf{R}^+$ describes the linear combination of the Spherical Harmonics that create a Hemispherical Harmonic base function. As an example, the columns of $\mathbf{R}^+$ with indices 5, 100 and 160 are presented in Figure 4.16 in dB with a dynamic range of 60 dB. The values are arranged by order and degree, as it is often done in plots of SH coefficients.

It can be seen that for the lower indices the base functions on the hemisphere are a combination of the subset of the Spherical Harmonics of the same degree m' (see Figure 4.16a with $m' = -2$ and Figure 4.16b with $m' = 9$). The base function $\hat{\underline{Y}}_{n'}^{m'}$ of order n' and degree m' can be calculated according to the following pattern regarding the columns of $\mathbf{R}^+$ and the Spherical Harmonics $\underline{Y}_n^{m'}$[4]:

$$\hat{\underline{Y}}_{n'}^{m'} = \sum_{n=n'}^{N_{\max}} \mathbf{R}^+(n^2 + n + m' + 1, n'^2 + n' + m' + 1) \cdot \underline{Y}_n^{m'}. \qquad (4.12)$$

Note that this behavior does not correspond to the case of symmetry with respect to the z-axis, which would be achieved by taking only the Spherical Harmonics where $n + m$ is an even number [84].

For high indices (and thus high orders), the strictly regular pattern concerning the degree m' can no longer be found. The 160th base function $\hat{\underline{Y}}_{12}^3$ (Figure 4.16c) is now also composed of Spherical Harmonics of different degrees and orders. This indicates that the conversion between $\hat{\underline{Y}}$ and the original Spherical Harmonics base could become unstable.

The truncation parameter for the SVD can be determined based on the expression in Eq. (4.12). By demanding that for each order and degree all entries in the columns of the matrix $\mathbf{R}^+$ must only have non-zero values at the indices related to the same degree, an optimization goal can be formulated:

$$t(n', m') = \sqrt{\frac{\displaystyle\sum_{n=n'}^{N_{\max}} \left| \mathbf{R}^+(n^2 + n + m' + 1, n'^2 + n' + m' + 1) \right|^2}{\displaystyle\sum_{n=1}^{N_{\max}} \sum_{m=-n}^{n} \left| \mathbf{R}^+(n^2 + n + m + 1, n'^2 + n' + m' + 1) \right|^2}} \qquad (4.13)$$

---

[4]Here, the indexing relation $idx(n, m) = n^2 + n + m + 1$ has been used, which relates the order and degree to the linear index in the base function matrix and coefficient vector

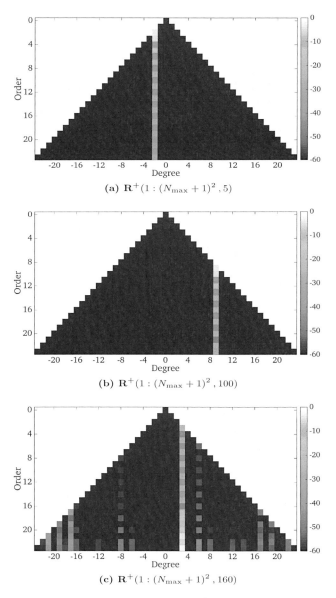

(a) $\mathbf{R}^+(1:(N_{\max}+1)^2,5)$

(b) $\mathbf{R}^+(1:(N_{\max}+1)^2,100)$

(c) $\mathbf{R}^+(1:(N_{\max}+1)^2,160)$

Figure 4.16: Columns of the matrix $\mathbf{R}^+$ relating the Hemispherical Harmonics to the Spherical Harmonics

This measure lies between zero and one. The optimization is then based on maximizing the order n' for which the value of $t(n', m')$ is above a certain threshold; in the case presented here the threshold was chosen as 0.8.

In Figure 4.17 the result of the optimization of the truncation parameter for the SVD is presented. Figure 4.17a shows what will from now on be called the *maximum stable order* for the base functions as a function of the truncation parameter. Additionally, the lower limit of the truncation parameter to achieve a maximum order of $n' = 11$ is indicated by the dotted line. As the truncation is performed so that all the singular values belonging to a certain order are kept, discrete steps in the output of the optimization function can be seen. This also means that the truncation parameter can be chosen as any value above $2 \cdot 10^{-3}$ and the same result is obtained concerning the maximum stable order. This of course only holds until the truncation parameter is so large that not enough singular values remain to create enough base functions.

For the optimized value of the truncation parameter of $2 \cdot 10^{-3}$, Figure 4.17b presents the values of $t(n', m')$ according to Eq. (4.13). It can be confirmed that for $n' \geq 12$ the values are significantly below one. It may hence be better to reduce the data to orders smaller than 12.

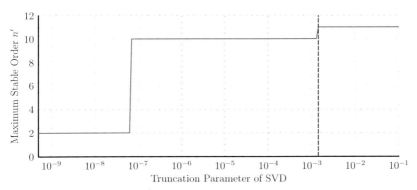

(a) Maximum stable order n' as a function of the truncation parameter for the SVD

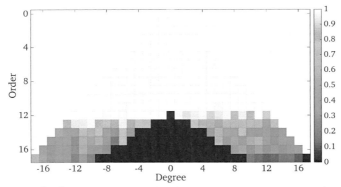

(b) $t(n', m')$ according to Eq. (4.13) for a truncation parameter of $2 \cdot 10^{-3}$

Figure 4.17: Optimization of the truncation parameter of the SVD for a maximum stable order of the base functions

B Analytical Examples

To test the approach of spatial filtering of data obtained on hemispherical shells, some analytical examples using monopole sources will be employed. Starting with a single monopole source outside of the array representing the loudspeaker, the complexity of the setup will be increased by adding an image source and a source inside of the array.

In this context, the sound pressure distribution at the microphones due to the point sources will be calculated both in the spatial domain using Eq. (2.17) as well as in the SH domain applying Eq. (2.31). This enables the identification of the effect of spatial aliasing, as each solution is correct in the respective domain. Hence by comparing the results between domains gives an impression of the influence of spatial aliasing. The calculation in the SH domain is of course only exact up to the order N_{max} used in Eq. (2.31).

As a measure of comparison of spatial data, the spatial cross-correlation $\underline{C}(\mathbf{p}_1, \mathbf{p}_2)$ between the (pressure) data vectors $\mathbf{p}_1$ and $\mathbf{p}_2$ will be employed [77]. Taking $\mathbf{p}_1$ as the reference, the normalized cross-correlation defined by

$$\underline{C}(\mathbf{p}_1, \mathbf{p}_2) = \frac{\mathbf{p}_1^H \mathbf{p}_2}{\|\mathbf{p}_1\|^2}, \tag{4.14}$$

compares both the shape as well as the amplitude and phase of the data, with values of one for perfect agreement. The definition of the cross-correlation in Eq. (4.14) is valid both in the spatial domain as well as the SH domain, working either directly on the pressure at the microphones or on the SH coefficients.

In the following investigations the setup as depicted in Figure 4.18 will be used. The distance of the sources outside of the array (denoted by S_1 and IS) to the array center is $2\,\mathrm{m}$ and the angle of incidence is $\theta_0 = 45\,°$. The source inside the array is located at $(x, y, z) = (-0.1, 0.2, 0.05)$ m.

In a first step it has to be validated that the original base functions can be used to correctly transform the sound field. This has been done by calculating the correlation according to Eq. (4.14) between the analytical result (Eq. (2.31)) and the SH transform using Eq. (2.29). As the spatial sampling of the individual arrays no longer follows the Gauss-Legendre quadrature rule, the pseudo-inverse of the SH matrix has been used for the transform. The resulting spatial correlation for each of the three sources as well as the superposition of all sources is presented in Figure 4.19a. Additionally, the spatial correlation between the original sound field and the inverse SH transform has been calculated. The result is shown inFigure 4.19b.

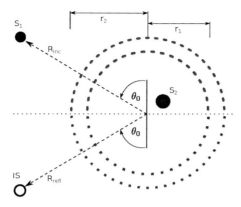

Figure 4.18: Schematic of the setup for the analytical investigations on the feasibility of the SH transform on the hemisphere

It can be confirmed that the SH transform on the full sphere can be successfully applied and that the inverse transform correctly reproduces the original spatial data as the correlation values are exactly equal to one up to the angular aliasing frequency of 2.2 kHz.

The effect of aliasing is not large for the SH coefficients of the individual source signals, where only a slight variation of the correlation can be found. However for the superposition of all three sources, larger deviations from the desired value of one occur. For the correlation in the spatial domain, spatial aliasing becomes even more evident as the values above 2.2 kHz drop drastically. This is not surprising as the inverse SH transform with the order $N_{max} = 23$ is basically a spatial low-pass filter. The correlation calculated for the source inside the array does not show any aliasing effects as the aliasing frequency for point sources closer to the origin (i. e. with a smaller radius) is higher (compare Eq. (4.11)).

Analogous to the case of the full sphere, the correlation between the analytical result and the transformed sound field has been calculated for the hemisphere, using the previously determined base functions (Figure 4.15) for the HSH transform. The analytical HSH coefficients have been converted from the full sphere with the transformation matrix according to Eq. (2.51). The results are presented for the HSH domain in Figure 4.20a and for the spatial domain in Figure 4.20b.

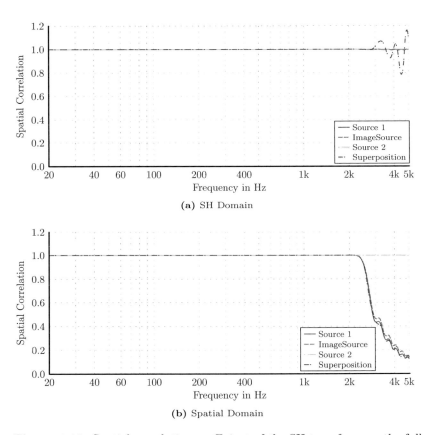

(a) SH Domain

(b) Spatial Domain

Figure 4.19: Spatial correlation coefficient of the SH transform on the full sphere for the setup depicted in Figure 4.18

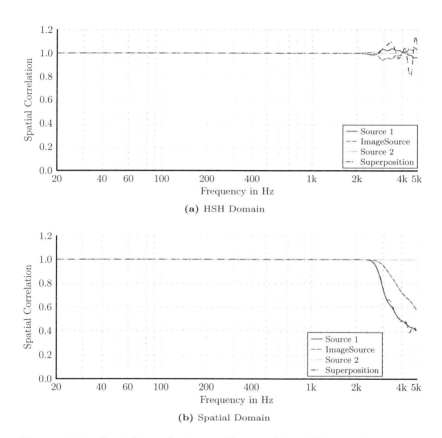

(a) HSH Domain

(b) Spatial Domain

Figure 4.20: Spatial correlation coefficient of the HSH transform on the hemisphere for the setup depicted in Figure 4.18

Very similar results to those obtained for the full sphere (Figure 4.19) can be observed for the hemisphere, with the difference that the effect of aliasing in the HSH domain (Figure 4.20a) is larger than in the SH domain (Figure 4.19a). The variation of the correlation values above 2.2 kHz is slightly larger in this case, which suggests increased susceptibility to aliasing errors.

In the spatial domain in Figure 4.20b the behavior of the correlation across frequency is similar to the case on the full sphere. However, the correlation for the image source (dashed curve) is higher than the corresponding curve in Figure 4.19b. This is not expected and an explanation cannot be given at this point.

Nevertheless it can be stated that the HSH transform works well up to the aliasing frequency. With this knowledge, the next step is to test the applicability of the spatial filtering methods in the HSH domain, i. e. beamforming and near-field holography as described in Section 2.2.4 and Section 2.2.3, respectively. This analysis will determine whether the radial filters can also be applied to the result of the HSH transform.

Again, in a first step the optimal result of the spatial filtering methods will be analyzed for the full sphere using the coefficients in the SH domain. Following the setup depicted in Figure 4.18, the source signal of the direct source S_1 was reconstructed in three situations: (1) only S_1 was active, (2) S_1 and the image source (IS) were active, and (3) S_1, IS and the source inside the array (S_2) were active. For a realistic scenario, the image source was used to model a reflection from an absorber. Hence, the signal for IS was that of S_1 multiplied by the reflection factor related to the impedance of the locally reacting material used in Section 2.3 (see dashed curve in Figure 2.7). Similarly, S_2 was used to model a reflection from inside the array with the same reflection factor and an additional loss in amplitude due to spherical wave propagation from S_1 to the array center.

In Figure 4.21 the original source signal together with the spatial filtering results is presented for the three scenarios described above. The filtering methods described in Section 2.2.3 and Section 2.2.4 have been applied. Additionally, beamforming has also been calculated in the spatial domain (denoted by $S2$).

It can be confirmed that in the ideal case of a single source (Figure 4.21a) the signal can be reconstructed perfectly in the aliasing-free frequency range. This is true for both beamforming as well as holography. With the image source additionally active (Figure 4.21b), the outgoing pressure determined by near-field holography is still zero, which is the expected behavior as the image source is also located outside of the array.

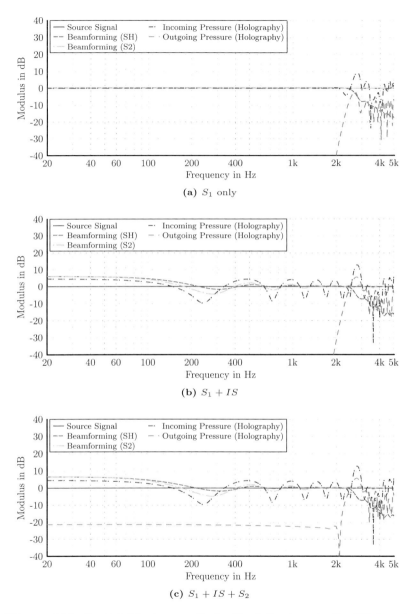

Figure 4.21: Spatial filtering results for the full sphere in the SH domain using the setup depicted in Figure 4.18

The incoming sound pressure now contains both the direct signal as well as the image source signal and hence does not directly yield the desired result.

The beamforming methods produce a good result starting from 200 Hz. For low frequencies the spatial selectivity is not high enough. *Plane-Wave Decomposition* (PWD), which has a higher directivity index at low frequencies, could also be used but this method is relatively unstable and the better directivity index comes at the cost of low noise rejection [81, 165]. Since noise is a problem especially at low frequencies, where the loudspeaker does not radiate sound as efficiently, PWD will not be used.

The beamforming result in the SH domain performs slightly better than in the spatial domain at frequencies between 200 Hz and 2 kHz. However, beamforming in the S2 domain is not affected by spatial aliasing as much and hence gives better results at frequencies above 2 kHz.

The performance of the beamforming methods is not considerably reduced by the additional source inside the array (Figure 4.22c), as the results are practically identical to the ones in Figure 4.21b. The holography method is able to correctly extract the outgoing wave signal. It can be concluded that both beamforming and holography perform well on the full sphere in the SH domain. For the desired application of determining the source reference signal, beamforming is better suited. As already mentioned before, the result of the holography method can additionally be analyzed with beamforming to separate the different incident signals but the result then does not differ much from the direct application of beamforming. This could change if the amplitude of the outgoing waves becomes stronger.

With the spatial filtering results in the SH domain as the reference, the same analysis has been performed in the HSH domain for the data on the hemisphere. The results are shown in Figure 4.22 in the same order and for the same filtering methods as in Figure 4.21.

From the data for the direct source in Figure 4.22a it can clearly be seen that near-field holography does not work as stable as it does in the SH domain. The outgoing sound pressure, which should in this case be zero, is only about 13 dB below the incoming pressure and at low frequencies the difference is even less than that.

It seems that the transformation of the coefficients from the HSH to the SH domain (see Eq. (2.53)) is not accurate enough to obtain a good holography result. Unfortunately, the order reduction mentioned in Section 4.2.4.A also does not lead to better results.

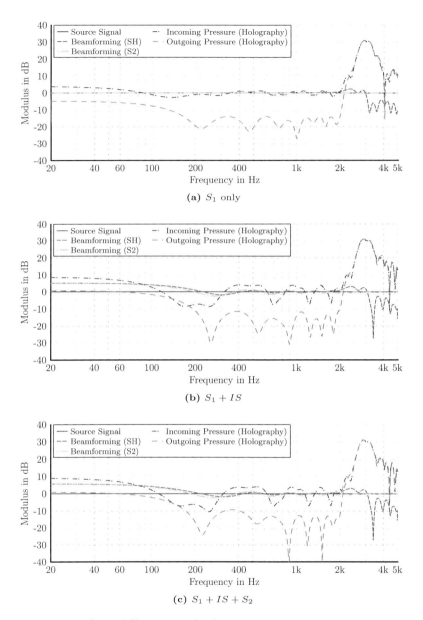

Figure 4.22: Spatial filtering results for the hemi-sphere in the HSH domain using the setup depicted in Figure 4.18

On the other hand, the application of beamforming in the HSH seems to work equally well as in the SH domain (compare dashed curves in Figure 4.21a and Figure 4.22a). The source signal can be reconstructed until the aliasing frequency. Just as before, beamforming in the spatial domain works well also above 2 kHz.

Similarly, for the case with the image source (Figure 4.22b) and the additional source inside the array (Figure 4.22c) the beamforming results are comparable to those obtained on the full sphere. The processing in the HSH domain yields slightly better results than the conventional beamforming in the spatial domain. The effective aliasing frequency in the HSH domain is reduced from 2.2 kHz to 2 kHz, which does not considerably limit the application.

In conclusion, it has to be stated that the method of near-field holography works well in the SH domain with data obtained on a full sphere, as was also found in [162]. However, the application to data in the HSH domain seems to fail, yielding unphysical data due to the transformation of the coefficients from the HSH to the SH domain. The application of beamforming works well in both the SH and HSH domain and hence is the preferred choice to obtain an *in-situ* reference for the measurements of reflection properties. Whether the accuracy is high enough to determine the reflection properties remains a topic for future research.

4.3 Analysis of Reflected Sound: Reflection Factor

Once the reflected sound $\underline{p}_{\text{refl}}(k)$ has been determined separately, either using the subtraction method or spatial filtering methods, the spherical reflection factor can be calculated by rearranging Eq. (2.72):

$$\underline{Q}(k, \theta) = \frac{\underline{p}_{\text{refl}}(k)}{\underline{G}(k, R_{\text{refl}})} . \tag{4.15}$$

If the spherical reflection factor has been determined from measurements according to Eq. (4.15), the surface impedance can be calculated based on Eq. (2.73). Unfortunately, the equation cannot be directly solved for $\underline{Z}_S$ and hence the surface impedance has to be found by numerical optimization or iterative methods [166]. In the context of direct measurements of the field impedance with pu-probes, this has been discussed by ALVAREZ and JACOBSEN [167].

The complicated procedure of an optimization approach can be avoided in those cases where the plane wave approximation holds, i.e. when $\underline{Q}(k, \theta) \approx \underline{R}(k, \theta)$. In the next section, this will be explored for different reflection models to establish rules that indicate which model to choose for the impedance deduction based on the position of the source and receiver above the boundary.

4.3.1 Sound Reflection Models

As has been mentioned in Section 2.3.4, the approximate image source model can be used instead of the Complex Image Source Model (CISM) if the sum of heights of the source and receiver above the reflecting plane is larger than several acoustic wavelengths (see Eq. (2.75)). In this section, this rule-of-thumb and the applicability to different models to estimate the surface impedance from Eq. (4.15) will be analyzed.

It should be noted at this point that the Sommerfeld Integral Solution (SIS) (Eq. (2.78)) yields exactly the same results as the CIS model but takes significantly longer to compute. It will hence not be considered any further here. Nonetheless it serves to validate that the CISM equations yield the correct result for locally reacting materials.

The analysis of the reflection models has been performed in the following way: first the spherical reflection factor according to Eq. (2.73) has been calculated for an infinite plane with a specific normalized surface impedance ζ_S. In the next step, the mentioned models were used to reconstruct the impedance as $\hat{\zeta}_S$. In the case of the plane wave reflection factor, this was easily achieved by inverting Eq. (2.58):

$$\hat{\underline{\zeta}}_S(k,\theta) \approx \frac{1}{\cos(\theta)} \cdot \frac{1 + \underline{Q}(k,\theta)}{1 - \underline{Q}(k,\theta)} \,. \tag{4.16}$$

For the Error Function Solution (EFS) model in Eq. (2.79), a nonlinear least-squares optimization (with the MATLAB function lsqnonlin) was used to find the impedance giving the best fit to the obtained values of $\underline{Q}$.

Calculations were carried out for angles of specular reflection between 0 degree and 80 degrees. The values of the sum of heights of source and receiver relative to the acoustic wavelength were varied from 0.01 to 1000. The results for the relative error of the plane wave approximation according to Eq. (4.16) are presented in Figure 4.23. The three values of $\underline{\zeta}_S$ that were chosen for the analysis are the same ones used by SUH and NELSON [47] and ARETZ [3].

157

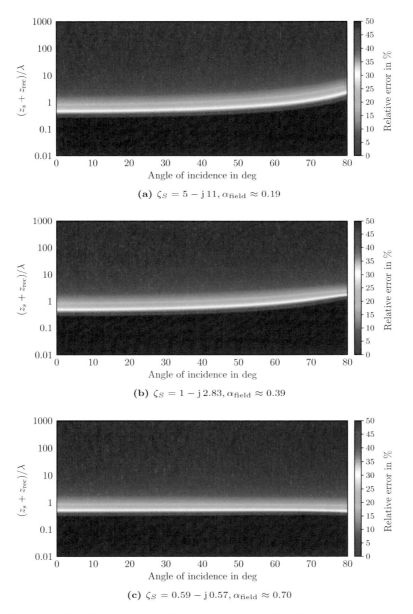

(a) $\zeta_S = 5 - \mathrm{j}\,11, \alpha_{\mathrm{field}} \approx 0.19$

(b) $\zeta_S = 1 - \mathrm{j}\,2.83, \alpha_{\mathrm{field}} \approx 0.39$

(c) $\zeta_S = 0.59 - \mathrm{j}\,0.57, \alpha_{\mathrm{field}} \approx 0.70$

Figure 4.23: Relative error of the estimated surface impedance in percent for the plane wave approximation according to Eq. (4.16)

The prediction according to Eq. (2.75) can be confirmed with the data shown in Figure 4.23. When the sum of heights is larger than approximately one wavelength the relative error decreases below 15 %. This is obviously not only the case for normal incidence, as was the supposition for the derivation of Eq. (2.75). A significant increase of the relative error can only be observed for angles of reflection greater than 40 degrees.

By comparing the results in Figure 4.23a and Figure 4.23b it can be observed that the error does not seem to depend on the impedance — and hence the absorption coefficient — for small angles of reflection. For high absorption as in Figure 4.23c the error actually decreases towards large angles of reflection becoming more or less constant.

An improved estimation of the surface impedance can be obtained if the property of spherical waves is accounted for in the derivation relating the reflection factor to the surface impedance in Eq. (2.58). The only difference is that the free-field impedance for spherical waves $\underline{Z}_f$ (see Eq. (2.19)) has to be incorporated instead of Z_0. Concerning the equation for the normalized surface impedance this leads to

$$\hat{\underline{\zeta}}'_S(k,\theta) \approx \frac{1}{\cos{(\theta)} \cdot \left(1 + \frac{1}{jk R_{\mathrm{spec}}}\right)} \cdot \frac{1 + \underline{Q}(k,\theta)}{1 - \underline{Q}(k,\theta)}, \tag{4.17}$$

where R_{spec} is the distance between the source and the point of specular reflection on the surface, i. e. the position where the straight line between the image source and the receiver intersects the surface (compare Figure 2.9b or Figure 4.8). The relative error for the modified plane wave approximation in Eq. (4.17) is presented in Figure 4.24 for the exact same scenarios as in Figure 4.23.

The data shows that the error can be significantly reduced for small angles of specular reflection, whereas for larger angles of reflection the error increases slightly. Concerning the dependence on the absorptive properties, similar observations as for the original plane wave model in Eq. (4.16) (see Figure 4.23) can be made, where the error at large angles of reflection decreases with increasing absorption.

It can thus be concluded for the plane wave approximation that the introduction of the spherical wave impedance gives a better accuracy especially towards normal incidence. The rule-of-thumb of $(z_s + z_{\mathrm{rec}})/\lambda > 1$ seems to be a good indicator when the plane wave approximation can be used to deduce the surface impedance from a measurement of the spherical reflection factor. For the modified model in Eq. (4.17) this approach ensures relative errors of less than 5 % for angles of specular reflection below 40 degrees.

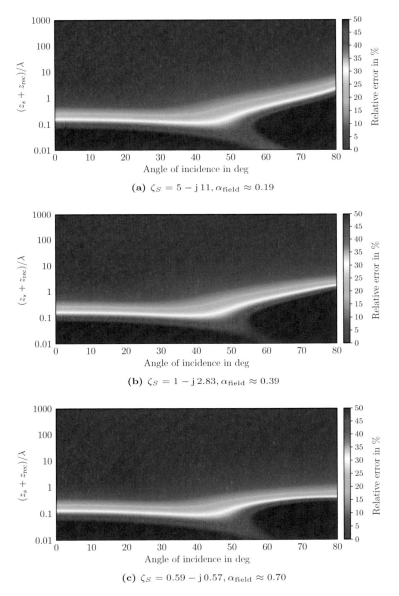

(a) $\zeta_S = 5 - \mathrm{j}\,11, \alpha_{\mathrm{field}} \approx 0.19$

(b) $\zeta_S = 1 - \mathrm{j}\,2.83, \alpha_{\mathrm{field}} \approx 0.39$

(c) $\zeta_S = 0.59 - \mathrm{j}\,0.57, \alpha_{\mathrm{field}} \approx 0.70$

Figure 4.24: Relative error of the estimated surface impedance in percent for the plane wave approximation including the spherical free-field impedance according to Eq. (4.17)

Another reflection model that was found to be practically applicable due to the relatively fast computation is the EFS model presented in Section 2.3.4. As already stated above this model cannot be inverted and hence the impedance has to be found from the obtained spherical reflection factor in an optimization process. This has been done for the scenario also used for the plane wave models above and the relative error with reference to the true value has been calculated. The results are shown in Figure 4.25.

In general, a far better approximation of the surface impedance can be observed compared to the plane wave models, especially for low values of the sum of heights relative to the acoustic wavelength. The EFS model was developed under the assumption of grazing (or at least near-grazing) angles of reflection and it can be confirmed from the data that the relative error decreases towards near-grazing angles of reflection. However, the model also produces good results for low angles of reflection.

The conclusion of the investigation of different reflection models is that for the sum of heights of source and receiver above the boundary significantly less than the acoustic wavelength only the CIS model can be used to correctly deduce the surface impedance. The solution involving the complementary error function gives good results for approximately $(z_{\mathrm{s}} + z_{\mathrm{rec}})/\lambda > 0.5$ and is computed efficiently, making it also applicable for an optimization approach. When source and receiver are more than one wavelength away from the boundary the error by using the plane wave approximation drops below $10\,\%$, especially when using the modified model in Eq. (4.17) including the spherical wave impedance.

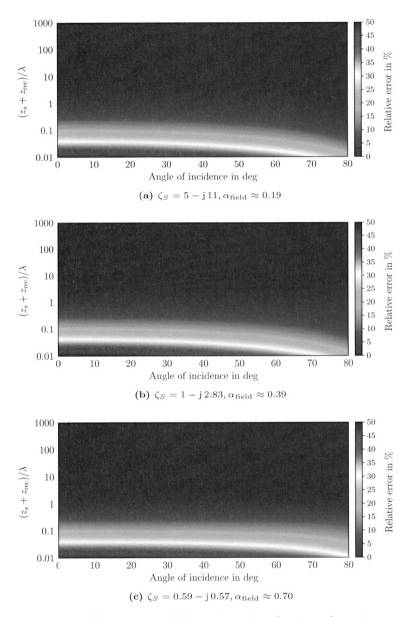

(a) $\zeta_S = 5 - j\,11, \alpha_{\text{field}} \approx 0.19$

(b) $\zeta_S = 1 - j\,2.83, \alpha_{\text{field}} \approx 0.39$

(c) $\zeta_S = 0.59 - j\,0.57, \alpha_{\text{field}} \approx 0.70$

Figure 4.25: Relative error of the estimated surface impedance in percent for the error function solution model

4.3.2 A Note on Models for Lateral Reaction

So far only models for locally reacting surfaces have been considered. The model by DI and GILBERT [95] for an infinitely extended and laterally reacting medium has been presented in Section 2.3.3. However, in the application of absorbing materials, which are usually backed by a sound-hard layer, this model may not be well suited. Another model specifically derived for the typical situation of a homogeneous porous absorber backed by an impervious layer has been introduced by ALLARD, LAURIKS, and VERHAEGEN [168]. Using this model for measurements above an absorber, the complex material parameters $\underline{\rho}_1$ and $\underline{k}_1$ could directly be obtained through curve-fitting.

A general model which relates the surface impedance of any material of arbitrary composition to the sound field above it has not been derived yet. Hence, in terms of flexible applicability, the best way to determine the reflection property of absorbing materials is to deduce the surface impedance using the models for local reaction mentioned in the last section. This has to be carried out for different angles of incidence to include the angle-dependency of the surface impedance. With this approach, the lateral reaction of a surface is considered at least with respect to the surface impedance, even if the exact sound field is not taken into account.

Further investigations on the topic of lateral reaction and the implications for measurement methods are definitely needed and important. However, since many free variables are involved — among them the flow resistivity and thickness of the porous material — such an investigation is outside the scope of this work. An investigation regarding laterally reacting materials and the described models has been carried out by DRAGONETTI and ROMANO [169].

4.3.3 Measurement Results

To verify that it is possible to obtain the desired data with the measurement setup described in Section 4.1, a series of measurements has been performed in the anechoic chamber of the ITA.[5] All measurements were performed at a sampling rate of 44.1 kHz in the frequency range between 100 Hz and 16 kHz. The length of the final impulse responses that were obtained using logarithmic sweeps [125] is 2048 samples $\hat{=}$ 46.4 ms. This leads to a frequency resolution of the final results of approximately 21.5 Hz.

---

[5] Due to a technical problem, not 24 but only 23 microphones could be used in the setup and hence the total number of measurement positions is 2208 instead of 2304. This basically means that the polar angle closest to 90 degrees is not covered.

163

Figure 4.26 presents the measurement setup to determine the reflection factor for three different scenarios:

- Figure 4.26a: the hardest test for reflection measurements is a surface with perfect reflection as has also been found in [66]. With an ideal reflection factor of one, this scenario can be used to determine the valid frequency range of the setup. The rigid floor in the anechoic chamber of the ITA has been measured as an example of a perfectly reflecting surface.

- Figure 4.26b: the opposite of the previous case is a perfectly absorbing surface, which has been realized with a large porous absorber with a flat surface. The effective thickness was $d = 250\,\mathrm{mm}$ and the flow resistivity of the material is $\Phi = 5.4\,\mathrm{kPa} \cdot \mathrm{s/m^2}$. The total area of the absorbing surface was $9\,\mathrm{m^2}$. Due to the relatively large extent of the absorber (more than one wavelength above 100 Hz), edge effects are not likely to influence the result. The reference for the acoustical impedance has been calculated with the optimized model by MIKI [90] with the mentioned material parameters.

- Figure 4.26c: a third example consists of a small sample of a porous absorber material with a flat surface. The effective thickness of this absorber is $d = 100\,\mathrm{mm}$ and the flow resistivity is assumed as $\Phi = 2.7\,\mathrm{kPa} \cdot \mathrm{s/m^2}$. The total area of the absorbing surface was $3\,\mathrm{m^2}$. The size of the absorber in this setup was relatively small (smaller than one wavelength below 250 Hz) so edge effects will more likely have an influence on the result. As in the previous case, the analytical reference has also been calculated with the optimized model by MIKI [90].

To reduce the group delay variation of the recorded signals all measured data has been deconvolved with the loudspeaker reference signal (Figure 4.3c). This makes the impulse responses more compact — in theory a Dirac impulse should be obtained. However, due to a limited *Signal-to-Noise Ratio* (SNR) for low frequencies where the loudspeaker does not efficiently radiate sound, only a band-limited deconvolution can be applied. The band-limitation slightly reduces the effectiveness of this method. Nevertheless, the approach can be used to enable additional time-windowing of the subtraction result.

For the results presented here the optimized subtraction method with a calibrated source as described in Section 4.2.2 has been applied. To obtain an impression of the performance of this approach in practical situations, Figure 4.27 shows typical results of the optimization process. In Figure 4.27a the optimal parameters of the (sub-)sample shift and the level change in Eq. (4.10) are presented for all microphone positions.

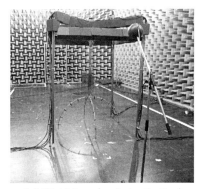

(a) Rigid floor (perfect reflection)

(b) Large flat absorber

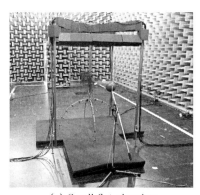

(c) Small flat absorber

Figure 4.26: Measurement scenarios to determine the reflection factor in the free-field

The resulting reduction in the signal energy — as the optimization goal — relative to the signal energy before subtraction is depicted in Figure 4.27b. The relative energy has been calculated for the direct application of the subtraction (solid curve), for the optimized subtraction (dashed curve) as well as for the result of the subsequent time-windowing (dash-dotted curve).

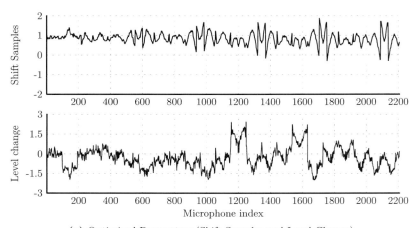

(a) Optimized Parameters (Shift Samples and Level Change)

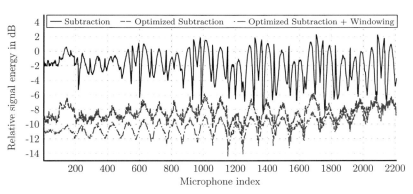

(b) Relative signal energy in dB compared to result before subtraction

Figure 4.27: Typical results of the optimization process for the subtraction method

From the optimized parameters in Figure 4.27a it can be seen that the equipment has been well calibrated and that the directivity model also seems to give correct results as the average change in level is close to 0 dB and rarely exceeds values of 1.5 dB. It can in this case be seen that the average number of shift samples is approximately equal to one, which means that the optimization concerning the location of loudspeaker and microphones described in Section 4.1.2 has not given the ideal result. Nevertheless, the boundary values of plus or minus two samples have not been reached confirming a correct setup.

The plots of the relative energy in Figure 4.27b show that the direct application of the subtraction method can in some cases even lead to an increase in level (values above 0 dB) due to a bad alignment of the recorded impulse response and the predicted direct impulse. The average improvement of the optimized subtraction method is 8 dB, which shows that it is an effective solution. From the fact that the optimized result never lies above the original one it can be seen that the optimization is stable. The subsequent time-windowing reduces the relative energy by another 2 dB for the receivers with low indices, which are further away from the reflecting surface. The difference of arrival for the receivers close to the boundary (with high indices) is too short to enable effective time-windowing so that there is no further improvement.

A comparison between the results without and with the directivity model presented in Section 4.1.4 has not shown large differences for the source and receiver positions used during the measurements presented here. This is because the maximum angle relative to the frontal direction of the loudspeaker never exceeds 30° so that there is not much level variation between the direct and reflected sound. Nevertheless, the proposed method works well also in this case. The directivity model becomes especially important when the source is very close to the receiver array or when the measurement setup spans a wider area and thus a larger solid angle relative to the frontal direction of the loudspeaker.

A Post-Processing of Measurement Data

From the measured data the spherical reflection factor is obtained according to Eq. (4.15) at many microphone positions and with a high frequency resolution. The high frequency resolution is not necessarily needed as the surface impedance of realistic materials does not vary strongly with frequency. In the application, most simulation tools only work with data in third-octave or even octave bands. Hence, a smoothing of the magnitude of the spherical

167

reflection factor has been applied to mitigate the influence especially of uncertainties in the low frequency range [170]. The smoothing bandwidth was chosen as ¹/₆-th of an octave.

The high number of microphone positions allows to obtain a more robust result by averaging the spherical reflection factor for all receivers with the same (or similar) angle of specular reflection. This has been performed by arranging the data in groups according to the angle of reflection with a quantization of 5 degrees. The effect of this averaging procedure can be understood assuming a ray-like behavior of sound reflection (see Figure 2.9b). By using the data at different receiver positions, an area-averaged reflection factor for the absorber surface is obtained because of the different positions of the specular reflection points on the surface; an illustration is given in Figure 4.28 for a specular reflection angle $\theta = 45\,^{\circ}$.

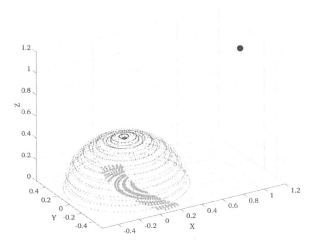

Figure 4.28: Schematic of the measurement setup and the implications of averaging results across microphones; source (blue), receivers (red) and specular reflections points on the surface (yellow) for a reflection angle $\theta = 45\,^{\circ}$

It can be seen that the reflection points on the surface span a wide area. The averaging procedure thus helps to reduce the effect of inhomogeneous material distributions and in part it also reduces the influence of edge diffraction as this effect is different at each receiver.

To remove reflections, e. g. from the measurement setup, a time-window has been applied for the high-frequency part of the result. The time-window was chosen such that signal components are removed that arrive 1.6 ms after the predicted impulse based on the position of the image source. The allowed additional time corresponds to a propagation distance of approximately 55 cm. Since the relatively short window duration leads to a significant loss of energy for low frequencies, the measured reflection factor without and with the time-window applied has been cross-faded in the frequency domain at 1.25 kHz.

To give an impression of the effect of the processing steps mentioned here, Figure 4.29 shows an example of a typical measurement result for an angle of specular reflection $\theta = 35°$. The measured spherical reflection factor is plotted as a function of frequency in Figure 4.29a as obtained with a single receiver without (dashed curve) and with time-windowing (dot-dashed curve). Additionally the smoothed result (dash-dotted curve) as well as the averaged result across multiple receivers with the same angle of reflection (solid curve) are shown. The resulting normalized surface impedance is shown in Figure 4.29b for the same cases and it has been calculated according to Eq. (4.17). For the source positions used here, the limiting frequency for the plane wave model related to the sum of heights of source and receiver is approximately 600 Hz. Below this frequency the surface impedance has been determined by least-squares optimization based on the exact CIS model.

The data in Figure 4.29a shows that the original result with a single microphone is subject to large variations. Especially at high frequencies above 1.5 kHz the effect of reflections from the equipment can be observed both in the magnitude as well as phase response. By applying the time-window and cross-fading the results in the frequency-domain the high-frequency result can be improved, yet still some variation remains. The additional smoothing operation helps to produce a better result at low frequencies. Finally, the averaging across several measurement positions with the same specular reflection angle yields the most stable result with the least variations across frequency.

The effect of the processing steps on the surface impedance in Figure 4.29b is not as large as for the reflection factor. Nonetheless, the relatively noisy original result (dashed curve) is converted into a very smooth signal with respect to frequency in both modulus and phase (solid curve).

In all following evaluations, all of the mentioned processing steps have been applied. Below 600 Hz the surface impedance has been determined based on the CIS model enforcing the minimum-phase property of the impedance. The result has then been converted back to the plane-wave reflection factor and absorption coefficient. The measurement results will cover angles of reflection between $\theta = 20°$ and $\theta = 55°$ as it has been suggested that these angles are most important for an assessment of the behavior for random-incidence [171].

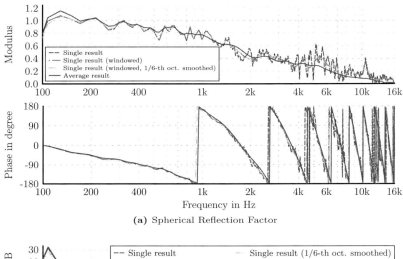

(a) Spherical Reflection Factor

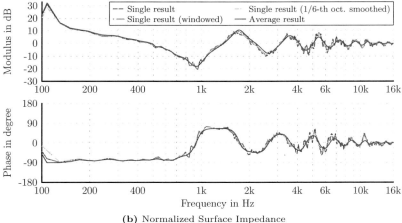

(b) Normalized Surface Impedance

Figure 4.29: Example of a measurement result of the spherical reflection factor and the resulting normalized surface impedance; angle of specular reflection is $\theta = 35\,^\circ$

As examples, single frequency responses will be presented for $\theta = 45\,^\circ$ which has been identified as a representative value for random sound incidence in Section 2.3.5.

B Rigid floor

As mentioned before, a first test on the quality of a measurement setup for the reflection factor is a sample that perfectly reflects all sound. Here, the rigid floor of the anechoic chamber at the ITA has been used for this purpose (see Figure 4.26a).

In Figure 4.30 the result of the complex plane-wave reflection factor is presented for measurements of the rigid floor and one source position, yielding results for angles of incidence between $\theta = 35\,^\circ$ and $\theta = 55\,^\circ$. The analytical reference of a reflection factor equal to one is additionally shown as the solid curve.

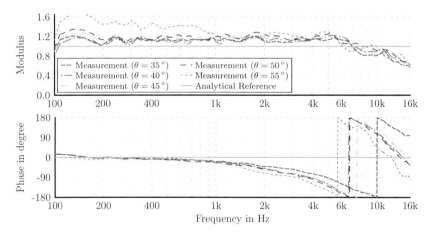

Figure 4.30: Measurement result of the plane-wave reflection factor of the rigid floor (see Figure 4.26a) for various angles of specular reflection

The data indicates that the modulus of the reflection factor can be determined between 100 Hz and approximately 6 kHz with a reasonable agreement to the reference. Above 6 kHz, the deviation from the correct value of one becomes larger, effectively predicting a higher absorption. It can also be observed that for an angle of reflection $\theta = 55\,^\circ$ especially at low frequencies the measured reflection factor is much higher than for the other angles. This can be related to the fact that for higher angles of specular reflection, not as many microphone positions are available for the averaging procedure described before. The lower values of the reflection factor for high frequencies suggest that it could be difficult to correctly measure samples with low absorption at high frequencies with the method described here.

It can also be seen that the phase deviates from the correct value starting from 1 kHz. The approximately linear negative slope of the phase suggests that the time of arrival for the image source has not been estimated correctly so that a slight time difference remains in the measured reflection factor. This shows that the results are very sensitive to the exact knowledge of the positions of the source and receivers. In the deduction of the surface impedance, errors in the phase response will have a strong effect, leading to a fluctuation of the modulus and phase of the surface impedance (compare Figure 4.29b). For a calculation of the absorption coefficient the phase of the reflection factor of course has no effect.

Due to remaining measurement uncertainties, the value of the reflection factor exceeds unity. In practical applications, such values should of course be truncated to one to prevent negative absorption coefficients. This is automatically achieved by enforcing the minimum-phase property when deducing the surface impedance.

C Large flat absorber

As an example of an absorber that can be modeled well by its thickness and flow resistivity using the model by MIKI [90], a large flat absorber made of polyurethane foam has been measured (see Figure 4.26b). The result of the complex plane-wave reflection factor for a specular reflection angle of $\theta = 45°$ is shown in Figure 4.31 (solid curve) together with the analytical result (dashed curve).

The modulus of the measured reflection factor agrees very well with the analytical calculation up to the maximum considered frequency of 16 kHz. There is a slight overestimation of the reflection factor between $300 - 600$ Hz. The phase response agrees relatively well with the analytical result up to 600 Hz, which is obviously due to the optimization process involving the CIS model. Above 600 Hz the trend of the phase response of the measured reflection factor seems to follow the analytical result but it is disturbed by fluctuations which can be observed by the many phase jumps. The importance of the phase for such a highly absorbing material is certainly questionable, as the value of the reflection factor is almost zero.

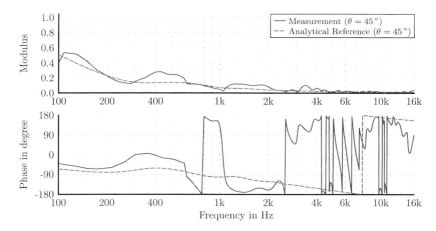

Figure 4.31: Measurement result of the plane-wave reflection factor of the large flat absorber (see Figure 4.26b) for an angle of specular reflection $\theta = 45\,°$

To obtain an impression on the performance of the measurement setup for different angles of specular reflection, the absorption coefficient has been evaluated from the measured data and it is shown in a contour plot as a function of frequency and reflection angle in Figure 4.32a. In addition, the analytical reference is presented in Figure 4.32b. The contour levels have been drawn in steps of 0.1.

In general, a fair agreement can be found between the analytical model and measurement. As stated before, the results for high angles of incidence are subject to larger variations which is due to few measurement positions to average across. For the absorber considered here, the absorption coefficient is equal to one throughout most of the frequency range and this can also be reproduced by the measurement method. Below 200 Hz, the absorption is overestimated by about 0.1 for most angles of reflection.

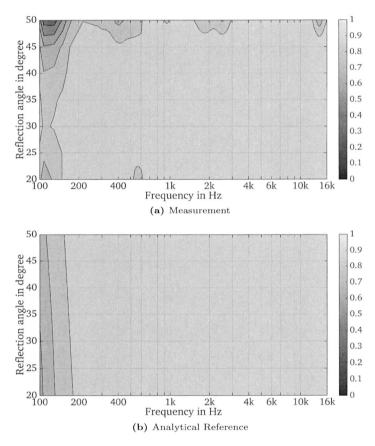

(a) Measurement

(b) Analytical Reference

Figure 4.32: Measurement result of the absorption coefficient of the large flat absorber (see Figure 4.26b) as a function of the specular reflection angle θ

D Small Flat Absorber

Another example of a porous absorber that has been measured is the small flat absorber depicted in Figure 4.26c. The measurement result of the plane-wave reflection factor for an angle of specular reflection of $\theta = 45\,^\circ$ is presented in Figure 4.33 (solid curve) together with the analytical result (dashed curve).

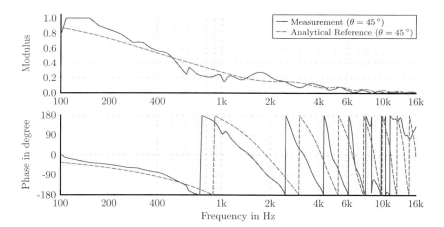

Figure 4.33: Measurement result of the plane-wave reflection factor of the small flat absorber (see Figure 4.26c) for an angle of specular reflection $\theta = 45\,^\circ$

As for the large absorber, the modulus of the measured and modeled reflection factor agrees well for frequencies above 1 kHz. Some variation across frequency can be observed which can be related to the edge effect due to the relatively small size of the sample. This has also been found in a previous numerical study of this measurement setup [147]. Larger deviations occur for frequencies between 500 Hz and 1 kHz, where the reflection factor is underestimated, and especially at low frequencies where it is overestimated.

The measured phase response follows the analytical one very well below 600 Hz. Above 600 Hz a similar effect as for the results for the rigid floor can be seen in so far as the measured response has a steeper slope of the phase indicating a time shift between measurement and model. This is again attributed to a slight mismatch between the estimated and the correct locations of source and receiver.

The measured and modeled absorption coefficient as a function of the specular reflection angle is shown in Figure 4.34a and Figure 4.34b, respectively.

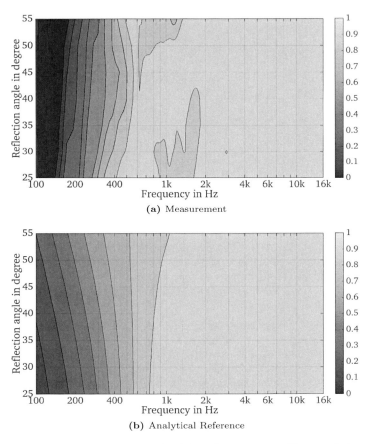

(a) Measurement

(b) Analytical Reference

Figure 4.34: Measurement result of the absorption coefficient of the small flat absorber (see Figure 4.26c) as a function of the specular reflection angle θ

The difference between measurement and analytical model is more obvious for the small absorber than for the large absorber. In the frequency range where the absorption of the sample is high, the agreement is good. However, for lower frequencies, where the absorption decreases, the error becomes larger and the absorption coefficient is underestimated. Below 200 Hz, the characteristic behavior with respect to the reflection angle is also not well captured by the measured data.

It can thus be concluded that the setup and the processing steps presented in this section can be used to obtain the absorbing properties of samples that are larger than the array diameter. The measurement results have shown that an imprecise knowledge of the source and receiver locations may result in errors in the phase response for the complex reflection factor which in turn leads to erroneous results of the surface impedance. It follows that a very careful setup of the measurement equipment is necessary.

4.4 Analysis of Reflected Sound: Spatial Response

An advantage of the setup with many microphones distributed on a hemisphere is that the reflected sound pressure can also be evaluated concerning the spatial response. This data can give additional insight into the process of sound reflection and it can eventually be used to determine the scattering properties of a surface. In this case, the directional diffusion coefficient (see Section 2.4.2) is an appropriate measure as the measurement setup according to the standards [116, 172] is very similar to the one presented here. Nonetheless, if a measurement with a flat reference surface can be performed, the directional scattering coefficient following the free-field correlation method as proposed by MOMMERTZ [113] can also be determined.

In this section, the application of the proposed setup regarding a measurement of the directional diffusion and scattering coefficient in the free-field is presented and discussed. The analysis of the spatial reflection data does not make much sense for the highly absorbing samples without a significant surface structure considered in Section 4.3. Thus, the sinusoid sample used for the verification measurements in Section 3.3.3.C has been used. The sample has a diameter of 800 mm, a profile peak-to-peak height of 20.4 mm and a structural wavelength of $\Lambda = 70.8$ mm.

To calculate the directional diffusion and scattering coefficient, the sound pressure that is scattered from a sample has to be determined. Usually, this is achieved by measuring the spatial distribution of the sound reflected from the sample under free-field conditions. The scattered sound pressure is then determined using the subtraction method [115]. With the setup envisioned here, the sample will be placed on a reflective ground, as shown in Figure 4.35 for the sinusoid sample mentioned before.

Figure 4.35: Setup for the measurement of the directional scattering properties in the anechoic chamber of the ITA

Based on the processing methods mentioned so far, two options exist to obtain the scattered sound pressure in this case:

1. **Near-Field Holography:** As already mentioned in Section 4.2.4, holographic methods are suitable to separate the incoming and outgoing waves with respect to the array. In the case of diffusion and scattering coefficient measurements, the outgoing waves (i. e. the coefficients $\underline{B}_{nm}$ in Eq. (2.36)) are then the desired quantity. In contrast to the application to absorbers larger than the array diameter, the necessary SH transform can be applied with the original base functions assuming symmetry with respect to the reflective ground. This prevents the numerical instabilities found in the application of Near-Field Holography using the Hemispherical Harmonics in Section 4.2.4.B. A major advantage of this approach is that the measurement environment does not need to be perfectly anechoic, as additional reflections from the surroundings can be separated with the spatial filtering methods.

2. **Subtraction Method:** If the source position and source reference signal are available, the subtraction method can be used to remove both the incident sound as well as the contribution by the image source, thus leaving only the sound pressure scattered by the sample. Although this method will probably be less reliable in practice due to the influence of uncertainties in the measurement setup (see Section 4.2.2), it effectively gives the same result as that obtained by the holography approach. As the combined array can be used in this case, the effective aliasing frequency is more than twice as high as for the holography approach.

Typically, free-field measurements of the diffusion and scattering coefficient are carried out in the far-field with a receiver distance of 5 m [116]. To obtain a comparable result with the method presented here, the scattered sound pressure can be extrapolated to the far-field by applying the appropriate Hankel function term in the SH domain [83] (compare Eq. (2.34)).

For an investigation of the proposed method under controlled conditions, numerical simulations with the Boundary-Element-Method (BEM) using *LMS Virtual.Lab 13.1* have been performed. The setup depicted in Figure 4.35 has been recreated by a CAD model of the sample and a point source representing the loudspeaker. The same source position as for the real measurement situation has been chosen, where the source was 2.3 m away from the array center at an angle of approximately 45 ° with respect to the surface normal. The direction of sound incidence was perpendicular to the surface corrugations.

Simulations have been carried out in the frequency range between 50 Hz and 10 kHz in steps of 50 Hz. To validate the results obtained with the array, additional simulations for a source distance of 10 m and a receiver distance of 5 m were performed to comply with the AES diffusion standard. The scattered sound pressure at the array microphones was extrapolated in the SH domain to the same receiver distance of 5 m. The diffusion coefficient was then calculated according to Eq. (2.85). The scattering coefficient was calculated based on the spatial cross-correlation as described in [113].

The results are shown in Figure 4.36a and Figure 4.36b for the diffusion and scattering coefficient, respectively, using the pressure obtained by Near-Field Holography (solid curve) and by the subtraction method (dashed curve). The result in the far-field is also shown as a reference (dot-dashed curve). For the scattering coefficient, the analytical solution calculated with the method proposed by Holford [32] using the implementation details given by EMBRECHTS et al. [35] is additionally plotted (dash-dotted curve).

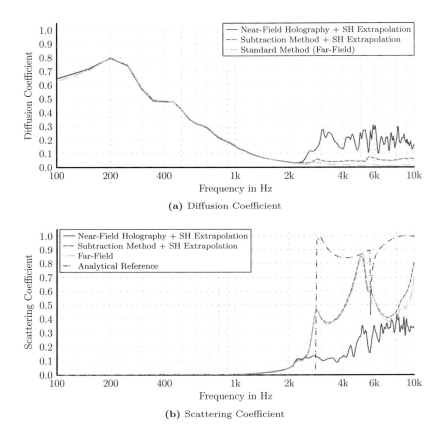

(a) Diffusion Coefficient

(b) Scattering Coefficient

Figure 4.36: Results of the directional diffusion and scattering coefficient of the sinusoid sample for an incident angle of $\theta = 45\,^\circ$ ($\varphi = 0\,^\circ$)

The curves for the diffusion coefficient in Figure 4.36a are practically identical up to the angular aliasing frequency of 2.2 kHz for the holography method (compare Figure 4.21). As the combined array has been used for the subtraction method, the array aliasing frequency lies higher at approximately 4.7 kHz. Due to the small sample size, the actual aliasing frequency for the scattered sound pressure is roughly 6 kHz. Yet obvious aliasing artifacts cannot be seen in the result for the diffusion coefficient. This is not entirely surprising, as the SH extrapolation to the far-field attenuates the high orders, which are most affected by aliasing. Still, an overestimation of the diffusion coefficient can be observed starting from 3 kHz.

The data for the scattering coefficient in Figure 4.36b shows that the relatively low aliasing frequency for the holography approach does not allow a meaningful measurement in this case, as the sample only scatters sound starting from approximately 2.8 kHz. Nonetheless, the results agree very well between the array methods and the far-field method up to 2.2 kHz. The subtraction method with a subsequent SH extrapolation yields basically the same scattering coefficient as the far-field method up to a frequency of 7 kHz. For higher frequencies there is again a slight overestimation as was also seen for the diffusion coefficient.

The comparison of the numerical data to the analytical result shows relatively poor agreement. The increase in the scattering coefficient occurs for the same frequency of approximately 2.8 kHz but is not as sudden as predicted by the theory. This has been noted in other studies [35]. The magnitude of the scattering coefficient is lower — sometimes by a factor of as much as two — compared to the analytical result. A dip can be observed for the analytical data at 5.8 kHz and the simulated results also show a drop in the scattering coefficient at approximately this frequency.

However, opposed to the analytical result, the numerically obtained scattering coefficient does not directly increase again for higher frequencies. This is not an effect of spatial aliasing, as simulations repeated with a higher resolution for the receiver array give similar results. A possible explanation for the discrepancies between the numerical and analytical result is that the analytical equations assume an infinite extent of the scattering surface whereas the sample is of finite extent in the numerical simulation. The evaluation method according to Mommertz used here should theoretically eliminate the influence of a finite surface by employing a flat reference sample with the same dimensions as the scattering sample. It is thus not exactly clear at this point where the deviations come from and this remains to be investigated in future research.

181

It can be concluded that the setup and following signal processing steps described in this section can be effectively used to obtain the directional scattering properties of samples that are smaller than the diameter of the array. Comparison to established far-field methods have shown very good agreement. A limit of the applicability lies in the size and resolution of the array. Since modern sensor technology allows to build arrays with hundreds of microphones, this is not a considerable challenge. In practical situations, a precise knowledge of the source and sensor positions is crucial, especially for the subtraction method, which has already been noted in Section 4.3.3.

5

Conclusion and Outlook

In this thesis, the topic of obtaining accurate boundary conditions related to the reflection of sound at architectural surfaces has been investigated. Thereby, both the absorbing as well as the scattering reflection properties have been considered. The measurement methods in the reverberation chamber as well as in the free-field have been treated separately.

5.1 Uncertainty Analysis of Reverberation Chamber Measurements

In the first part of the thesis, the standardized measurement methods in the reverberation chamber have been analyzed concerning measurement uncertainties. The method of error propagation has been applied to the analytical formulations relating the input quantities such as room volume, sample surface area and reverberation time to the absorption and scattering coefficient. Through an analysis of the developed equations, the most important factors that lead to increased uncertainties have been identified and discussed. The uncertainty of the room volume and sample surface area was found to be low enough to neglect the impact on the absorption and scattering coefficient.

Concerning systematic errors, the simplified equations in ISO 354 [5] to calculate the absorption coefficient were found to have a significant influence. The simplification to disregard the correction for the room surface area covered by the sample is potentially only a problem for scattering coefficient measurements. By using the Sabine equation instead of the Eyring equation, large errors can be found especially when measuring the scattering coefficient. As expected, the error becomes larger for higher absorption coefficients of the empty chamber and of the sample.

A definite answer to the question whether Sabine's or Eyring's equation is "correct" — if that is possible at all — has still not been given. Experimentally, this is difficult to achieve, because too many uncontrollable uncertainty factors remain. More importantly, prediction models for the absorption coefficient are not accurate enough to be used as a ground truth. The analytical models for scattering coefficients may be a better alternative and first comparisons have shown very good agreement [37]. It would be important to include the absorption coefficient of the sample material into the analytical formulas to achieve a better prediction. Numerical simulations could also be used, especially since nowadays advanced algorithms exist that allow calculations of very large problems.

Concerning stochastic errors, the theoretical predictions by DAVY [140] have been used to investigate the influence of random spatial fluctuations of the reverberation times in more detail. Very good agreement was found for the verification with measurement data, especially after including the low-frequency correction related to modal overlap. Based on the theoretical predictions, a more general analysis has been performed relating the measurement uncertainty to the dimensions and absorbing properties of the reverberation chamber and the sample. Relatively simple equations were derived to determine the minimum number of source-receiver combinations to achieve a given measurement precision. In the case of absorption coefficient measurements, the ISO recommendation of 12 measurement positions yields acceptable results as long as $\alpha_s \leq 0.5$ at low frequencies.

For measurements of the scattering coefficient it was confirmed that the absorption coefficient of the sample should be as low as possible because it dramatically increases the measurement uncertainty. The ISO limit of $\alpha_s = 0.5$ may lead to extremely high uncertainties if the sample scatters sound at low frequencies. In any case, the necessary number of measurement positions to achieve a given uncertainty is always larger for the scattering coefficient than for the absorption coefficient. Additionally, it was found that correlation between the measured reverberation times has to be considered when predicting the uncertainty of the scattering coefficient. Thus, it is recommended to change the corresponding equation in Annex A of ISO 17497-1 to include this effect.

Currently, an ISO working group is dealing with the revision of ISO 354 with the specific goal to achieve a better reproducibility, especially across laboratories. First recommendations on how this could be achieved have been given by VERCAMMEN [173]. Since it has also been suggested to use the predictions by Davy as a reference concerning the diffuseness of the measurement chamber, further investigations on this topic are necessary. Basically, a round-robin test on the accuracy of the uncertainty predictions could be carried out. The low-frequency term connected to the modal overlap is of special importance, as no reliable data exists for different laboratories.

The method for a prediction of the necessary number of source-receiver combinations developed in this thesis could be used as a more accurate estimate in ISO 354 (which is also referenced in ISO 17497-1). This would assure that the measurement uncertainty in each laboratory is more or less similar, which would make it easier to compare data between laboratories.

5.2 Measurement of Angle-Dependent Reflection Properties

In the second part of the thesis, the free-field measurement of angle-dependent reflection properties has been treated. A setup of a sequential hemispherical microphone array has been introduced that enables the measurement of both the absorption as well as the scattering coefficients in less than 20 minutes per source position. The influence of the array support structure on the sound field was determined experimentally and then minimized based on results in different configurations.

Regarding the measurement of the absorbing properties, an extension to the subtraction method has been developed to include the source and receiver directivity. Since the source used in the setup does not show a very pronounced directivity for small angles relative to the frontal direction, the effect was almost not noticeable in the measurement results presented in this thesis. Nonetheless, the approach can be applied easily and it has been shown in other studies that the improvement can be significant [155].

As an alternative to the subtraction method, array-processing techniques have been investigated. For the application above large absorbers, Hemispherical Harmonics (HSH) base functions have been derived and applied to analytical simulations. The use of beamforming in the HSH domain has shown equally good results as in the Spherical Harmonics (SH) domain. Results for Near-Field Scattering Holography on the hemisphere indicate that the conversion between the HSH and the SH domain leads to inaccuracies that prohibit a successful application. In the SH domain, however, the holographic method can be employed and it is especially useful for measurements of the scattering properties with the restriction that the sample has to be smaller than the diameter of the array.

The use of beamforming in the SH or in the spatial domain to obtain an *in-situ* reference signal has to be validated. For the data obtained in this thesis, relatively good performance has been seen. However, the question whether the accuracy is high enough to be applicable to the determination of reflection factors remains unanswered. The observed problems in an application of holography in the HSH domain should be investigated further, since the method has proven to be very robust in the SH domain and could outperform the beamforming methods.

Models of sound reflection for locally reacting materials have been investigated for a deduction of the surface impedance from the measured reflection factor. It was found that the simplified plane-wave model can only be used when the sum of source and receiver heights above the reflecting surface is larger than several wavelengths. The model involving the complimentary error function gives better results, however only the model based on image-sources with complex positions yields correct results at low frequencies for sources and receivers close to the surface.

Measurements regarding the absorbing properties in three different scenarios have been conducted. The results show that the introduced setup can be used to measure the absorption coefficient of highly absorbing samples. By comparing the results to analytical model calculations, it has been shown that even for samples of finite size good results can be obtained. Uncertainties in the knowledge of the source and receiver positions lead to errors in the phase response of the complex reflection factor. These errors also affect the subsequent deduction of the surface impedance using the sound reflection models.

Numerical simulations for a small scattering surface have been carried out to test the ability of the introduced array setup to measure the directional diffusion and scattering coefficient. Very good agreement was found between the array-processing methods with a subsequent extrapolation of the scattered sound in the SH domain and established far-field methods. In the case of the scattering coefficient, reasonable agreement was also found with analytical results.

The remaining uncertainties in the positions of the source and receivers should be reduced. A possibility would be to permanently integrate the microphones into the array support, so that the degrees of freedom are reduced. With today's options concerning 3D printing and MEMS microphone technology, this should pose no problem. An automatic tracking of the source, e. g. with an optical tracking system, would make the measurement with many different positions — and hence angles of incidence — easier and more robust.

The array-processing is limited to low and medium frequencies due to the spatial resolution. Because of the design of the sequential array, only the polar angle resolution has to be increased with more physical microphones, as the azimuthal resolution can be freely chosen. Again referring to the low-cost MEMS technology, building an array with many more sensors is certainly possible. It should be noted, however, that MEMS microphones typically have a much lower SNR and show more variation in the frequency response than electret capsules.

With an improved version of the introduced setup, the *in-situ* application should be tested. Such a measurement scenario poses new challenges regarding additional room reflections, which could be overcome by the array-processing methods. In fact, by using near-field holography, the room reflections could be used as secondary source signals to simultaneously measure at various angles of incidence.

A subject that has not been addressed in this work is the measurement of small samples with considerable influence of the edges and the surroundings on the effective absorption. Based on the work by THOMASSON [174], variational formulations have been used to describe the relation between the size of an absorber and its effective absorption coefficient [175, 176]. The array setup could be used to contribute to this research by measuring absorbing samples smaller than the array diameter. This would also make the signal processing more stable.

Acknowledgments

This dissertation and the years leading toward it have been influenced by important people in my life and it is now time to thank all of them for providing support and motivation throughout these past years.

First of all I would like to thank Prof. Michael Vorländer, not only for supervising the thesis, but for having provided me with the opportunity to do my Ph.D. at the ITA. The freedom granted to me and all colleagues to do research is exceptional and it is one of the reasons why such great work has been done at the ITA. I would also like to thank Prof. Jean-Jacques Embrechts from the Université de Liège for taking the time to be the second reviewer and for providing helpful comments on the dissertation.

Special thanks go to Dr. Gottfried Behler who not only gave me the first job at the ITA as a student nine years ago but who has since then been a constant source of knowledge and support. I am very grateful for the many fruitful discussions, be it concerning one of the many projects we worked on together, or any other topic.

I have worked on many measurement setups that would have never overcome the planning stage if it hadn't been for the great support of the mechanical and electrical workshops at the ITA, providing technical expertise and creative opportunities that are very rare. Representative for all colleagues in the workshops, I would like to thank Uwe Schlömer and Rolf Kaldenbach for their support.

All my colleagues at the ITA have to be acknowledged, as it is thanks to them that the working environment is so very special, regarding diverse acoustical knowledge as well as motivation and friendship. From the first years special thanks go to Dr. Dirk Schröder, Dr. Pascal Dietrich, Dr. Bruno Masiero and Dr. Martin Pollow. During the recent years especially Sönke Pelzer, Rob Opdam, Martin Guski and Johannes Klein have been great colleagues. I am also very grateful to all former students who put a lot of effort and work into their projects, which directly or indirectly aided with this thesis.

During the last years I was fortunate to spend time with some great musicians. Being in the studio and on the road with the bands helped a great deal to take the mind off work. I would like to thank my band colleagues Anja, Armin, Christian and Volker for this and for all the fun we had together.

Most importantly, I would like to express my deepest gratitude towards my mother. Her support and encouragement to pursue my path have helped me get to this point. Even if it did not always seem like it, I am forever thankful to her. Also, although they could not live to see this, I would like to thank my grandparents who have always generously supported me. I am also very thankful to my godmother Petra who has been a support to me and my mother in all the important moments in my life.

Last but definitely not least I want to thank my girlfriend Heike for her love and continuing faith in me. There is so much I could not have done without her in the last years and I am glad to have her by my side.

Bibliography

[1]   M. VORLÄNDER. *Auralization – Fundamentals of Acoustics, Modelling, Simulation, Algorithms and Acoustic Virtual Reality.* Springer Berlin, 2007.

[2]   D. SCHRÖDER. "Physically base real-time auralization of interactive virtual environments". PhD thesis. Institute of Technical Acoustics, RWTH Aachen University, 2011.

[3]   M. ARETZ. "Combined wave- and ray-based room acoustic simulations in small rooms". PhD thesis. Institute of Technical Acoustics, RWTH Aachen University, 2012.

[4]   W. C. SABINE. *Collected Papers on Acoustics.* Harvard University Press, Cambridge, 1923, p. 290.

[5]   *ISO 354: Acoustics – Measurement of sound absorption in a reverberation room.* ISO, 2003.

[6]   *ASTM C423: Acoustics – Standard test method for sound absorption and sound absorption coefficients by the reverberation room method.* ASTM, 2009.

[7]   *ISO 17497-1: Acoustics – Sound-scattering properties of surfaces – Part 1: Measurement of the random-incidence scattering coefficient in a reverberation room.* ISO, 2004.

[8]   U. KATH and W. KUHL. "Einfluss von Streufläche und Hallraumdimensionen auf den gemessenen Schallabsorptionsgrad (Influence of the scattering surface and reverberation chamber dimensions on the measured sound absorption coefficient)". In: *Acustica* 11 (1961), pp. 50–64.

[9]   E. TOYODA, S. SAKAMOTO, and H. TACHIBANA. "Effects of room shape and diffusing treatment on the measurement of sound absorption coefficient in a reverberation room". In: *Acoustical Science and Technology* 25.4 (2004), pp. 255–266.

[10] A. COPS, J. VANHAECHT, and K. LEPPENS. "Sound absorption in a reverberation room: Causes of discrepancies on measurement results". In: *Applied Acoustics* 46.3 (1995). Building Acoustics, pp. 215 –232.

[11] A. C. C. WARNOCK. "Some practical aspects of absorption measurements in reverberation rooms". In: *The Journal of the Acoustical Society of America* 74.5 (1983), pp. 1422–1432.

[12] W. KUHL. "Der Einfluß der Kanten auf die Schallabsorption poröser Materialien (The influence of edges on the sound absorption of porous materials)". In: *Acustica* 10 (1960), pp. 264–276.

[13] W. KUHL. "Ursachen und Verhinderung systematischer Abweichungen vom wahren Absorptionsgrad bei der Absorptionsgradmessung im Hallraum". In: *Acta Acustica united with Acustica* 52.4 (1983), pp. 197–210.

[14] A. de BRUIJN. "A Mathematical Analysis Concerning the Edge Effect of Sound Absorbing Materials". In: *Acta Acustica united with Acustica* 28.1 (1973), pp. 33–44.

[15] T. W. BARTEL. "Effect of absorber geometry on apparent absorption coefficients as measured in a reverberation chamber". In: *The Journal of the Acoustical Society of America* 69.4 (1981), pp. 1065–1074.

[16] C. G. BALACHANDRAN. "Random Sound Field in Reverberation Chambers". In: *The Journal of the Acoustical Society of America* 31.10 (1959), pp. 1319–1321.

[17] G. VENZKE and P. DÄMMIG. "Measurement of Diffuseness in Reverberation Chambers with Absorbing Material". In: *The Journal of the Acoustical Society of America* 33.12 (1961), pp. 1687–1689.

[18] R. V. WATERHOUSE. "Statistical Properties of Reverberant Sound Fields". In: *The Journal of the Acoustical Society of America* 43.6 (1968), pp. 1436–1444.

[19] T. F. W. EMBLETON. "Absorption Coefficients of Surfaces Calculated from Decaying Sound Fields". In: *The Journal of the Acoustical Society of America* 49.1A (1971), pp. 99–99.

[20] T. J. SCHULTZ. "Diffusion in reverberation rooms". In: *Journal of Sound and Vibration* 16.1 (1971), pp. 17–28.

[21] H. G. ANDRES and D. BRODHUN. "Zur Genauigkeit von Schallabsorptionsgradsmessungen im Hallraum (On the accuracy of sound absorption coefficient measurements in the reverberation chamber)". In: *Acustica* 10 (1960), pp. 330–335.

[22] E. MEYER and H. KUTTRUFF. "Akustische Modellversuche zum Aufbau eines Hallraums (Acoustical model experiments on the setup of

a reverberation chamber)". In: *Nachr. Akad. Wiss. Göttingen* 6.1958 (1958), pp. 97–114.

[23]   C. W. KOSTEN. "International comparison measurements in the reverberation room". In: *Acustica* 10 (1960), pp. 400–411.

[24]   Y. MAKITA, M. KOYASU, M. NAGATA, and S. KIMURA. "Investigations into the precision of measurement of sound absorption coefficients in a reverberation room (II), Experimental studies on the method of measurement of the reverberation time and the 4th round robin test". In: *J. Acoust. Soc. Japan,(in Japanese)(Australia, CSIRO Translation (English), No. 10821)* 24 (1968), pp. 393–402.

[25]   R. OHLAN. *Nordic comparison measurements of absorption coefficient.* SP-RAPP 13. Swedish National Testing and Research Institute, 1977.

[26]   H. MYNCKE, D. COPS, and D. DEVRIES. "The measurement of the sound absorption coefficient in reverberation rooms and results of a recent round robin test". In: *Third Symposium of the Federation of Acoustical Societies of Europe, Yugoslavia.* 1979, pp. 259–272.

[27]   W. A. DAVERN and P. DUBOUT. *First report on Australasian comparison measurements of sound absorption coefficients.* Commonwealth Scientific and Industrial Research Organization, Division of Building Research, 1980.

[28]   R. E. HALLIWELL. "Inter-laboratory variability of sound absorption measurement". In: *The Journal of the Acoustical Society of America* 73.3 (1983), pp. 880–886.

[29]   M. VORLÄNDER and E. MOMMERTZ. "Definition and measurement of random-incidence scattering coefficients". In: *Applied Acoustics* 60.2 (2000), pp. 187 –199.

[30]   M. GOMES, M. VORLÄNDER, and S. GERGES. "Aspects of the sample geometry in the measurement of the random-incidence scattering coefficient". In: *Proc. Forum Acusticum Sevilla.* 2002.

[31]   Y.-J. CHOI and D.-U. JEONG. "Some Issues in Measurement of the Random-Incidence Scattering Coefficients in a Reverberation Room". In: *Acta Acustica united with Acustica* 94.5 (2008), pp. 769–773.

[32]   R. L. HOLFORD. "Scattering of sound waves at a periodic, pressure-release surface: An exact solution". In: *The Journal of the Acoustical Society of America* 70.4 (1981), pp. 1116–1128.

[33] J.-J. EMBRECHTS, D. ARCHAMBEAU, and G. STAN. "Determination of the Scattering Coefficient of Random Rough Diffusing Surfaces for Room Acoustics Applications". In: *Acta Acustica united with Acustica* 87.4 (2001), pp. 482–494.

[34] J.-J. EMBRECHTS and A. BILLON. "Theoretical Determination of the Random-Incidence Scattering Coefficients of Infinite Rigid Surfaces with a Periodic Rectangular Roughness Profile". In: *Acta Acustica united with Acustica* 97.4 (2011), pp. 607–617.

[35] J.-J. EMBRECHTS, L. DE GEETERE, G. VERMEIR, M. VORLÄNDER, and T. SAKUMA. "Calculation of the Random-Incidence Scattering Coefficients of a Sine-Shaped Surface". In: *Acta Acustica united with Acustica* 92.4 (2006), pp. 593–603.

[36] M. VORLÄNDER, J.-J. EMBRECHTS, L. DE GEETERE, G. VERMEIR, and M. H. de AVELAR GOMES. "Case Studies in Measurement of Random Incidence Scattering Coefficients". In: *Acta Acustica united with Acustica* 90.5 (2004), pp. 858–867.

[37] I. SCHMICH-YAMANE, J.-J. EMBRECHTS, M. MÜLLER-TRAPET, C. ROUGIER, M. MALGRANGE, and M. VORLÄNDER. "Prediction and measurement of the random-incidence scattering coefficient of periodic reflective rectangular diffuser profiles". In: *Proceedings of Meetings on Acoustics (POMA): ICA 2013, Montreal, Canada.* 2013, 015144, 9 p.

[38] S. J. KLINE and F. A. McCLINTOCK. "Describing uncertainties in single-sample experiments". In: *Mechanical Engineering* 75.1 (1953), pp. 3–8.

[39] R. J. MOFFAT. "Describing the uncertainties in experimental results". In: *Experimental Thermal and Fluid Science* 1.1 (1988), pp. 3 –17.

[40] A. LUNDEBY, T. E. VIGRAN, H. BIETZ, and M. VORLÄNDER. "Uncertainties of Measurements in Room Acoustics". In: *Acta Acustica united with Acustica* 81.4 (1995), pp. 344–355.

[41] M. GUSKI and M. VORLÄNDER. "Comparison of Noise Compensation Methods for Room Acoustic Impulse Response Evaluations". In: *Acta Acustica united with Acustica* 100.2 (2014), pp. 320–327.

[42] P. DÄMMIG and H. DEICKE. "Measurement uncertainty in the determination of the sound absorption in a reverberation room at low frequencies". In: *Acustica* 33 (1975), pp. 249–256.

[43]   H. H. Ku. "Notes on the use of propagation of error formulas". In: *Journal of Research of the National Bureau of Standards* 70.4 (1966), pp. 263–273.

[44]   *Evaluation of measurement data – Guide to the expression of uncertainty in measurement.* BIPM, 2008.

[45]   V. Wittstock. "On the Uncertainty of Single-Number Quantities for Rating Airborne Sound Insulation". In: *Acta Acustica united with Acustica* 93.3 (May 2007), pp. 375–386.

[46]   M. Vorländer. "Computer simulations in room acoustics: Concepts and uncertainties". In: *The Journal of the Acoustical Society of America* 133.3 (2013), pp. 1203–1213.

[47]   J. S. Suh and P. A. Nelson. "Measurement of transient response of rooms and comparison with geometrical acoustic models". In: *Journal of the Acoustical Society of America* 105.4 (1999), pp. 2304–2317.

[48]   C.-H. Jeong, J.-G. Ih, and J. H. Rindel. "An approximate treatment of reflection coefficient in the phased beamtracing method for the simulation of enclosed sound fields at medium frequencies". In: *Applied Acoustics* 69 (2008), pp. 601–613.

[49]   C.-H. Jeong, D. Lee, S. Santurette, and J.-G. Ih. "Influence of impedance phase angle on sound pressures and reverberation times in a rectangular room". In: *The Journal of the Acoustical Society of America* 135.2 (2014), pp. 712–723.

[50]   *ISO 10534-2: Acoustics – Determination of sound absorption coefficient and impedance in impedance tubes – Part 2: Transfer-function method.* ISO, 1998.

[51]   S. Pelzer, M. Müller-Trapet, and M. Vorländer. "Angle-dependent reflection factors applied in room acoustics simulation". In: *40th annual congress of AIA and 39th German Annual Conference on Acoustics (DAGA), Merano, Italy.* 2013, pp. 1649–1652.

[52]   L. de Geetere. "Analysis and improvements of the experimental techniques to assess the acoustical reflection properties of boundary surfaces". PhD thesis. KU Leuven, 2004.

[53]   E. Mommertz. "Angle-dependent in-situ measurements of reflection coefficients using a subtraction technique". In: *Applied Acoustics* 46.3 (1995), pp. 251–263.

[54] J.-F. ALLARD and B. SIEBEN. "Measurements of acoustic impedance in a free field with two microphones and a spectrum analyzer". In: *The Journal of the Acoustical Society of America* 77.4 (1985), pp. 1617–1618.

[55] J.-F. ALLARD and Y. CHAMPOUX. "In Situ Two-Microphone Technique for the Measurement of the Acoustic Surface Impedance of Materials". In: *Noise Control Engineering Journal* 32.1 (1989), pp. 15–23.

[56] M MINTEN, A. COPS, and W. LAURIKS. "Absorption characteristics of an acoustic material at oblique incidence measured with the two-microphone technique". In: *Journal of sound and vibration* 120.3 (1988), pp. 499–510.

[57] M. TAMURA. "Spatial Fourier transform method of measuring reflection coefficients at oblique incidence. I: Theory and numerical examples". In: *The Journal of the Acoustical Society of America* 88.5 (1990), pp. 2259–2264.

[58] M. TAMURA, J.-F. ALLARD, and D. LAFARGE. "Spatial Fourier-transform method for measuring reflection coefficients at oblique incidence. II. Experimental results". In: *The Journal of the Acoustical Society of America* 97.4 (1995), pp. 2255–2262.

[59] T. BARRY. "Measurement of the absorption spectrum using correlation/spectral density techniques". In: *The Journal of the Acoustical Society of America* 55 (1974), p. 1349.

[60] J. C. DAVIES and K. A. MULHOLLAND. "An impulse method of measuring normal impedance at oblique incidence". In: *Journal of Sound and Vibration* 67.1 (1979), pp. 135–149.

[61] J. S. BOLTON and E. GOLD. "The application of Cepstral techniques to the measurement of transfer functions and acoustical reflection coefficients". In: *Journal of Sound Vibration* 93 (1984), pp. 217–233.

[62] M. GARAI. "Measurement of the sound-absorption coefficient in situ: The reflection method using periodic pseudo-random sequences of maximum length". In: *Applied Acoustics* 39 (1993), pp. 119 –139.

[63] R. LANOYE, G. VERMEIR, W. LAURIKS, R. KRUSE, and V. MELLERT. "Measuring the free field acoustic impedance and absorption coefficient of sound absorbing materials with a combined particle velocity-pressure sensor". In: *The Journal of the Acoustical Society of America* 119.5 (2006), pp. 2826–2831.

[64] H.-E. de BREE, P. LEUSSINK, T. KORTHORST, H. JANSEN, T. LAMMERINK, and M. ELWENSPOEK. "The μ-flown: a novel device for measuring acoustic flows". In: *Sensors and Actuators A: Physical* 54.1–3 (1996), pp. 552–557.

[65] E. BRANDÃO, R. C. C. FLESCH, A. LENZI, and C. A. FLESCH. "Estimation of pressure-particle velocity impedance measurement uncertainty using the Monte-Carlo method". In: *The Journal of the Acoustical Society of America* 130.1 (2011), EL25–EL31.

[66] M. MÜLLER-TRAPET, P. DIETRICH, M. ARETZ, M. van GEMMEREN, and M. VORLÄNDER. "On the in situ impedance measurement with pu-probes—Simulation of the measurement setup". In: *The Journal of the Acoustical Society of America* 134.2 (2013), pp. 1082–1089.

[67] T. J. HARGREAVES, T. J. COX, Y. W. LAM, and P. D'ANTONIO. "Surface diffusion coefficients for room acoustics: Free-field measures". In: *The Journal of the Acoustical Society of America* 108.4 (2000), pp. 1710–1720.

[68] I SCHMICH and N BROUSSE. "In Situ Measurement Methods for Characterising Sound Diffusion". In: *Building Acoustics* 18.1 (2011), pp. 17–36.

[69] M. KLEINER, H. GUSTAFSSON, and J. BACKMAN. "Measurement of Directional Scattering Coefficients Using Near-Field Acoustic Holography and Spatial Transformation of Sound Fields". In: *Audio Engineering Society Convention 99*. Oct. 1995.

[70] H. KUTTRUFF. *Acoustics*. Taylor & Francis, 2007.

[71] L. L. BERANEK. *Acoustics*. Published by the American Institute of Physics for the Acoustical Society of America, 1986.

[72] *ISO 2533: Standard-Atmosphere*. ISO, 1975.

[73] H. E. BASS, L. C. SUTHERLAND, A. J. ZUCKERWAR, D. T. BLACKSTOCK, and D. M. HESTER. "Atmospheric absorption of sound: Further developments". In: *The Journal of the Acoustical Society of America* 97.1 (1995), pp. 680–683.

[74] *ISO 9613-1: Acoustics – Attenuation of sound during propagation outdoors – Part 1: Calculation of the absorption of sound by the atmosphere*. ISO, 1993.

[75] E. WILLIAMS. *Fourier Acoustics: Sound Radiation and Nearfield Acoustical Holography*. Academic Press, 1999.

[76] F. ZOTTER. "Analysis and Synthesis of Sound-Radiation with Spherical Arrays". PhD thesis. University of Music and Performing Arts, Graz, Austria, 2009.

[77] M. POLLOW. "Directivity patterns for room acoustical measurements and simulations". PhD thesis. Institute of Technical Acoustics, RWTH Aachen University, 2014.

[78] F. ZOTTER. "Sampling strategies for acoustic holography/holophony on the sphere". In: *Meeting of the Dutch Acoustical Socitey (NAG) and 35th German Annual Conference on Acoustics (DAGA), Rotterdam, The Netherlands*. 2009.

[79] G. WEINREICH and E. B. ARNOLD. "Method for measuring acoustic radiation fields". In: *The Journal of the Acoustical Society of America* 68.2 (1980), pp. 404–411.

[80] H. VAN TREES. *Optimum Array Processing: Part IV of Detection, Estimation and Modulation Theory*. John Wiley & Sons, 2002.

[81] B. RAFAELY. "Phase-mode versus delay-and-sum spherical microphone array processing". In: *Signal Processing Letters, IEEE* 12.10 (Oct. 2005), pp. 713 –716.

[82] D. JOHNSON and D. DUDGEON. *Array Signal Processing: Concepts and Techniques*. Prentice Hall, 1993.

[83] M. POLLOW, K.-V. NGUYEN, O. WARUSFEL, T. CARPENTIER, M. MÜLLER-TRAPET, M. VORLÄNDER, and M. NOISTERNIG. "Calculation of Head-Related Transfer Functions for Arbitrary Field Points Using Spherical Harmonics Decomposition". In: *Acta Acustica united with Acustica* 98.1 (2012), pp. 72–82.

[84] H. POMBERGER and F. ZOTTER. "An ambisonics format for flexible playback layouts". In: *1st Ambisonics Symposium, Graz, Austria*. 2009.

[85] N. SNEEUW. "Global spherical harmonic analysis by least-squares and numerical quadrature methods in historical perspective". In: *Geophysical Journal International* 118.3 (1994), pp. 707–716.

[86] R. PAIL, G. PLANK, and W.-D. SCHUH. "Spatially restricted data distributions on the sphere: the method of orthonormalized functions and applications". English. In: *Journal of Geodesy* 75.1 (2001), pp. 44–56.

[87]   F. P. MECHEL. *Schallabsorber. 1. Äußere Schallfelder, Wechselwirkungen (Sound absorbers. Part 1: Exterior sound fields, interactions).* Hirzel, 1989.

[88]   A. V. OPPENHEIM, R. W. SCHAFER, and J. R. BUCK. *Discrete-Time Signal Processing.* 2nd ed. Prentice-Hall, Englewood Cliffs, NJ, 1999.

[89]   M. E. DELANY and E. N. BAZLEY. "Acoustical properties of fibrous absorbent materials". In: *Applied Acoustics* 3.2 (1970), pp. 105 –116.

[90]   Y. MIKI. "Acoustical properties of porous materials-Modifications of Delany-Bazley models". In: *Journal of the Acoustical Society of Japan (E)* 11.1 (1990), pp. 19–24.

[91]   T. KOMATSU. "Improvement of the Delany-Bazley and Miki models for fibrous sound-absorbing materials". In: *Acoustical Science and Technology* 29.2 (2008), pp. 121–129.

[92]   I. LINDELL and E. ALANEN. "Exact image theory for the Sommerfeld half-space problem, part I: Vertical magnetic dipole". In: *IEEE Transactions on Antennas and Propagation* 32.2 (Feb. 1984), pp. 126–133.

[93]   I. LINDELL and E. ALANEN. "Exact image theory for the Sommerfeld half-space problem, part II: Vertical electric dipole". In: *IEEE Transactions on Antennas and Propagation* 32.8 (Aug. 1984), pp. 841–847.

[94]   I. LINDELL and E. ALANEN. "Exact image theory for the Sommerfeld half-space problem, part III: General formulation". In: *IEEE Transactions on Antennas and Propagation* 32.10 (Oct. 1984), pp. 1027 –1032.

[95]   X. DI and K. E. GILBERT. "An exact Laplace transform formulation for a point source above a ground surface". In: *The Journal of the Acoustical Society of America* 93.2 (1993), pp. 714–720.

[96]   M. OCHMANN. "The complex equivalent source method for sound propagation over an impedance plane". In: *The Journal of the Acoustical Society of America* 116.6 (2004), pp. 3304–3311.

[97]   M. OCHMANN. "Closed form solutions for the acoustical impulse response over a masslike or an absorbing plane". In: *The Journal of the Acoustical Society of America* 129.6 (2011), pp. 3502–3512.

[98]   J. B. ALLEN and D. A. BERKLEY. "Image method for efficiently simulating small-room acoustics". In: *The Journal of the Acoustical Society of America* 65.4 (1979), pp. 943–950.

[99]   F. P. MECHEL. "Improved Mirror Source Method in Roomacoustics". In: *Journal of Sound and Vibration* 256.5 (2002), pp. 873 –940.

[100] A. SOMMERFELD. "Über die Ausbreitung der Wellen in der drahtlosen Telegraphie, On wave propagation for radio telegraphy". In: *Annalen der Physik* 28 (1909), pp. 665–736.

[101] U. INGARD. "On the Reflection of a Spherical Sound Wave from an Infinite Plane". In: *The Journal of the Acoustical Society of America* 23.3 (1951), pp. 329–335.

[102] R. B. LAWHEAD and I. RUDNICK. "Acoustic Wave Propagation Along a Constant Normal Impedance Boundary". In: *The Journal of the Acoustical Society of America* 23.5 (1951), pp. 546–549.

[103] C. CHIEN and W. SOROKA. "Sound propagation along an impedance plane". In: *Journal of Sound and Vibration* 43.1 (1975), pp. 9 –20.

[104] C. NOCKE, V. MELLERT, T. WATERS-FULLER, K. ATTENBOROUGH, and K. LI. "Impedance deduction from broad-band, point-source measurements at grazing incidence". In: *Acta Acustica united with Acustica* 83.6 (1997), pp. 1085–1090.

[105] M. R. STINSON. "A note on the use of an approximate formula to predict sound fields above an impedance plane due to a point source". In: *The Journal of the Acoustical Society of America* 98.3 (1995), pp. 1810–1812.

[106] E. T. PARIS. "On the coefficient of sound-absorption measured by the reverberation method". In: *The London, Edinburgh, and Dublin Philosophical Magazine and Journal of Science* 5.29 (1928), pp. 489–497.

[107] Y. MAKITA and T. HIDAKA. "Revision of the cos θ Law of Oblique Incident Sound Energy and Modification of the Fundamental Formulations in Geometrical Acoustics in Accordance with the Revised Law". In: *Acta Acustica united with Acustica* 63.3 (1987), pp. 163–173.

[108] Y. MAKITA and T. HIDAKA. "Comparison between Reverberant and Random Incident Sound Absorption Coefficients of a Homogeneous and Isotropic Sound Absorbing Porous Material – Experimental Examination of the Validity of the Revised cos θ Law". In: *Acta Acustica united with Acustica* 66.4 (1988), pp. 214–220.

[109] C. HOPKINS. *Sound Insulation*. Butterworth-Heinemann, 2007.

[110] M. ARETZ and M. VORLÄNDER. "Efficient Modelling of Absorbing Boundaries in Room Acoustic FE Simulations". In: *Acta Acustica united with Acustica* 96.6 (2010), pp. 1042–1050.

[111] *ISO 140-5: Acoustics – Measurement of sound insularion in bbuilding and of building elements – Part 5: Field measurement of airborne sound insulation of facade elements and facades.* ISO.

[112] H. KUTTRUFF. *Room acoustics.* Taylor & Francis, 2000.

[113] E. MOMMERTZ. "Determination of scattering coefficients from the reflection directivity of architectural surfaces". In: *Applied Acoustics* 60.2 (2000), pp. 201–203.

[114] F. L. PEDROTTI and L. S. PEDROTTI. *Introduction to Optics, 2nd Edition.* Vol. 1. Prentice Hall, 1993.

[115] T. COX and P. D'ANTONIO. *Acoustic absorbers and diffusers: theory, design and application.* Taylor & Francis, 2009.

[116] *AES-4id-2001: AES information document for room acoustics and sound reinforcement systems – characterisation and measurement of surface scattering uniformity.* AES, 2001.

[117] M. R. SCHROEDER. "Die statistischen Parameter der Frequenzkurve von großen Räumen (Statistical Parameters of the Frequency Response Curves of Large Rooms)". In: *Acustica* 4 (1954), 594—600.

[118] M. R. SCHROEDER and K. H. KUTTRUFF. "On Frequency Response Curves in Rooms. Comparison of Experimental, Theoretical, and Monte Carlo Results for the Average Frequency Spacing between Maxima". In: *The Journal of the Acoustical Society of America* 34.1 (1962), pp. 76–80.

[119] C. KOSTEN. "The mean free path in room acoustics". In: *Acustica* 10 (1960), pp. 245–250.

[120] H. KUTTRUFF. "Weglängenverteilung in Räumen mit schallzer-streuenden Elementen (Path length distribution in rooms with sound scattering elements)". In: *Acustica* 24 (1971), pp. 356–258.

[121] L. L. BERANEK and N. NISHIHARA. "Mean-free-paths in concert and chamber music halls and the correct method for calibrating dodecahedral sound sources". In: *The Journal of the Acoustical Society of America* 135.1 (2014), pp. 223–230.

[122] C. F. EYRING. "Reverberation time in "dead" rooms". In: *The Journal of the Acoustical Society of America* 1.2A (1930), pp. 217–241.

[123] I. N. BRONSHTEIN, K. A. SEMENDYAYEV, G. MUSIOL, and H. MUEHLIG. *Handbook of mathematics*. Vol. 3. Springer, 2007.

[124] J. BORISH and J. B. ANGELL. "An Efficient Algorithm for Measuring the Impulse Response Using Pseudorandom Noise". In: *Journal of the Audio Engineering Society* 31.7/8 (1983), pp. 478–488.

[125] S. MÜLLER and P. MASSARANI. "Transfer-Function Measurement With Sweeps". In: *Journal of the Audio Engineering Society* 49.6 (2001), pp. 443–471.

[126] *ISO 18233: Acoustics – Application of new measurement methods in building and room acoustics*. ISO, 2006.

[127] *ISO 10140-2: Acoustics – Laboratory measurement of sound insulation of building elements – Part 2: Measurement of airborne sound insulation*. ISO, 2010.

[128] M. R. SCHROEDER. "New Method of Measuring Reverberation Time". In: *The Journal of the Acoustical Society of America* 37.3 (1965), pp. 409–412.

[129] D. T. BRADLEY, M. MÜLLER-TRAPET, J. ADELGREN, and M. VORLÄNDER. "Effect of boundary diffusers in a reverberation chamber: Standardized diffuse field quantifiers". In: *The Journal of the Acoustical Society of America* 135.4 (2014), pp. 1898–1906.

[130] *ISO Guide 98: Uncertainty of measurement*. ISO, 2009.

[131] M. MÜLLER-TRAPET and M. VORLÄNDER. "Einfluss der Sabine'schen Näherungsformel für den Absorptionsgrad auf die Berechnung des Streugrades (Influence of Sabine's approximate equation for the absorption coefficient on the calculation of the scattering coefficient)". In: *37th German Annual Conference on Acoustics (DAGA), Düsseldorf, Germany*. 2011.

[132] M. MÜLLER-TRAPET and M. VORLÄNDER. "Uncertainty factors in determining the random-incidence scattering coefficient". In: *Proceedings of the 162nd Meeting of the Acoustical Society of America, San Diego, CA, USA*. Vol. 130. 4. Melville, NY: Acoustical Society of America, 2011, pp. 2355–2355.

[133] M. MÜLLER-TRAPET and M. VORLÄNDER. "Unsicherheitsfaktoren bei der Bestimmung des Streugrades nach der Hallraummethode (Uncertainty factors regarding the determination of scattering coefficients with the reverberation chamber method)". In: *38th German Annual Conference on Acoustics (DAGA), Darmstadt, Germany*. 2012.

[134]   M. MÜLLER-TRAPET and M. VORLÄNDER. "Uncertainty analysis of standardized measurements of random-incidence absorption and scattering coefficients". In: *The Journal of the Acoustical Society of America* 137.1 (2015), pp. 63–74.

[135]   SENSIRION. *Humidity and Temperature Sensors.* http://www.sensirion.com, last accessed April, 28th, 2015.

[136]   J. L. DAVY, I. P. DUNN, and P. DUBOUT. "The variance of decay rates in reverberation rooms". In: *Acustica* 43.1 (1979), pp. 12–25.

[137]   J. L. DAVY. "The variance of impulse decays". In: *Acta Acustica united with Acustica* 44.1 (1980), pp. 51–56.

[138]   J. L. DAVY and I. P. DUNN. "The statistical bandwidth of Butterworth filters". In: *Journal of Sound and Vibration* 115.3 (1987), pp. 539 –549.

[139]   *IEC 61260: Electroacoustics – Octave-band and fractional-octave-band filters.* IEC, 1995.

[140]   J. L. DAVY. "The variance of decay rates at low frequencies". In: *Applied Acoustics* 23.1 (1988), pp. 63 –79.

[141]   J. L. DAVY. "The relative variance of the transmission function of a reverberation room". In: *Journal of Sound and Vibration* 77.4 (1981), pp. 455 –479.

[142]   M. HODGSON. "Experimental evaluation of the accuracy of the Sabine and Eyring theories in the case of non-low surface absorption". In: *The Journal of the Acoustical Society of America* 94.2 (1993), pp. 835–840.

[143]   M. MÜLLER-TRAPET, R. VITALE, and M. VORLÄNDER. "A Revised Scale-Model Reverberation Chamber for Measurements of Scattering Coefficients". In: *2nd Pan-American/Iberian Meeting on Acoustics, Cancún, Mexico.* JASA, Nov. 2010.

[144]   T. SAKUMA, Y. KOSAKA, L. DE GEETERE, and M. VORLÄNDER. "Relationship Between the Scattering Coefficients Determined with Coherent Averaging and with Directivity Correlation". In: *Acta Acustica united with Acustica* 95.4 (2009), pp. 669–677.

[145]   M. MÜLLER-TRAPET and M. VORLÄNDER. "In-Situ Measurements of Surface Reflection Properties". In: *Building Acoustics* 21.2 (2014), pp. 167–174.

[146]   G. ISENBERG. *Design and Evaluation of an Acoustic Measurement Setup for the Determination of Material- and Angle- Dependent*

Reflection Properties. Bachelor Thesis, Institute of Technical Acoustics, RWTH Aachen University. 2012.

[147] J. MENDE. *Uncertainty analysis of an in-situ setup to measure the angle-dependent and complex reflection factor.* Bachelor Thesis, Institute of Technical Acoustics, RWTH Aachen University. 2014.

[148] J. RATHSAM and B. RAFAELY. "Analysis of in-situ acoustic absorption using a spherical microphone array". In: *Proceedings of Meetings on Acoustics* 6.1, 015002 (2009), pp. –.

[149] B. RAFAELY, B. WEISS, and E. BACHMAT. "Spatial Aliasing in Spherical Microphone Arrays". In: *Signal Processing, IEEE Transactions on* 55.3 (Mar. 2007), pp. 1003 –1010.

[150] N. THIELE. "Loudspeakers in Vented Boxes: Part 1". In: *Journal of the Audio Engineering Society* 19.5 (1971), pp. 382–392.

[151] N. THIELE. "Loudspeakers in Vented Boxes: Part 2". In: *Journal of the Audio Engineering Society* 19.6 (1971), pp. 471–483.

[152] R. H. SMALL. "Direct Radiator Loudspeaker System Analysis". In: *Journal of the Audio Engineering Society* 20.5 (1972), pp. 383–395.

[153] M. MÜLLER-TRAPET, P. DIETRICH, and M. VORLÄNDER. "Influence of Various Uncertainty Factors on the Result of Beamforming Measurements". In: *Noise Control Engineering Journal* 59.3 (2011), pp. 302–310.

[154] M. MÜLLER-TRAPET and M. VORLÄNDER. "Signal processing for hemispherical measurement data". In: *Proceedings of Meetings on Acoustics (POMA): ICA 2013, Montreal, Canada.* 2013, 055083, 9 p.

[155] M. MÜLLER-TRAPET and M. VORLÄNDER. "Berücksichtigung der Quellenrichtcharakteristik bei der *in-situ* Messung von Absorptionsgraden (Considering the source directivity in *in-situ* measurements of absorption coefficients)". In: *40th German Annual Conference on Acoustics (DAGA), Oldenburg, Germany.* 2014, pp. 232–233.

[156] H. BRAREN. *Using superposed acoustic monopoles to describe real sound source directivities to be used in analytic and numeric applications.* Bachelor Thesis, Institute of Technical Acoustics, RWTH Aachen University. 2013.

[157] J. ESCOLANO, J. J. LÓPEZ, and B. PUEO. "Directive sources in acoustic discrete-time domain simulations based on directivity

diagrams". In: *The Journal of the Acoustical Society of America* 121.6 (2007), EL256–EL262.

[158] R. DURAISWAMI, D. ZOTKIN, and N. GUMEROV. "Interpolation and range extrapolation of HRTFs [head related transfer functions]". In: *Acoustics, Speech, and Signal Processing, 2004. Proceedings. (ICASSP '04). IEEE International Conference on.* Vol. 4. May 2004, iv–45 –iv–48 vol.4.

[159] M. YUZAWA. "A method of obtaining the oblique incident sound absorption coefficient through an on-the-spot measurement". In: *Applied Acoustics* 8.1 (1975), pp. 27 –41.

[160] P. ROBINSON and N. XIANG. "On the subtraction method for in-situ reflection and diffusion coefficient measurements". In: *The Journal of the Acoustical Society of America* 127.3 (2010), EL99–EL104.

[161] F. DIERKES. *Angle-dependent reflection factors: Measurement methods and application in auralization.* Master Thesis, Institute of Technical Acoustics, RWTH Aachen University. 2012.

[162] M. MÜLLER-TRAPET, M. POLLOW, and M. VORLÄNDER. "Application of Scattering Nearfield Holography to In-Situ Measurements of Sound Absorption". In: *40th annual congress of AIA and 39th German Annual Conference on Acoustics (DAGA), Merano, Italy.* 2013.

[163] M. MÜLLER-TRAPET and M. VORLÄNDER. "Hemi-Spherical Harmonics – Verarbeitung von Messdaten auf Halbkugelschalen (Hemi-Spherical Harmonics – Processing of measurement data on hemispherical shells)". In: *41st German Annual Conference on Acoustics (DAGA), Nürnberg, Germany.* 2015.

[164] M. MÜLLER-TRAPET, M. POLLOW, and M. VORLÄNDER. "Spherical Harmonics as a Basis for Quantifying Scattering and Diffusing Objects". In: *Forum Acusticum – Aalborg, Denmark.* 2011.

[165] B. RAFAELY. "Plane-wave decomposition of the sound field on a sphere by spherical convolution". In: *Acoustical Society of America Journal* 116 (2004), pp. 2149–2157.

[166] E. BRANDÃO, E. TIJS, A. LENZI, and H.-E. de BREE. "A Comparison of Three Methods to Calculate the Surface Impedance and Absorption Coefficient from Measurements Under Free Field or in situ Conditions". In: *Acta Acustica united with Acustica* 97.6 (2011), pp. 1025–1033.

[167] J. D. ALVAREZ and F. JACOBSEN. "An Iterative Method for Determining the Surface Impedance of Acoustic Materials In Situ". In: *Internoise 2008, Shanghai, China*. 2008.

[168] J.-F. ALLARD, W. LAURIKS, and C. VERHAEGEN. "The acoustic sound field above a porous layer and the estimation of the acoustic surface impedance from free-field measurements". In: *The Journal of the Acoustical Society of America* 91.5 (1992), pp. 3057–3060.

[169] R. DRAGONETTI and R. A. ROMANO. "Considerations on the sound absorption of non locally reacting porous layers". In: *Applied Acoustics* 87.0 (2015), pp. 46 –56.

[170] T. G. H. BASTEN and H.-E. de BREE. "Full bandwidth calibration procedure for acoustic probes containing a pressure and particle velocity sensor". In: *The Journal of the Acoustical Society of America* 127.1 (2010), pp. 264–270.

[171] C.-H. JEONG and J. BRUNSKOG. "The equivalent incidence angle for porous absorbers backed by a hard surface". In: *The Journal of the Acoustical Society of America* 134.6 (2013), pp. 4590–4598.

[172] *ISO 17497-2: Acoustics – Sound-scattering properties of surfaces – Part 2: Measurement of the directional diffusion coefficient in a free field*. ISO.

[173] M. L. S. VERCAMMEN. "Improving the accuracy of sound absorption measurement according to ISO 354". In: *Proceedings of the International Symposium on Room Acoustics, Melbourne, Australia*. 2010.

[174] S.-I. THOMASSON. "On the absorption coefficient". In: *Acta Acustica united with Acustica* 44.4 (1980), pp. 265–273.

[175] F. MECHEL. "On sound absorption of finite-size absorbers in relation to their radiation impedance". In: *Journal of Sound and Vibration* 135.2 (1989), pp. 225 –262.

[176] D. HOLMBERG, P. HAMMER, and E. NILSSON. "Absorption and Radiation Impedance of Finite Absorbing Patches". In: *Acta Acustica united with Acustica* 89.3 (2003), pp. 406–415.

Curriculum Vitae

Personal Data

	Markus Müller-Trapet
17.09.1981	born in Düsseldorf, Germany

Education

1993–2001	Mataré-Gymnasium, Meerbusch-Büderich, Germany
1992–1993	Landrat-Lucas-Gymnasium, Leverkusen, Germany
1988–1992	GGS Theodor-Fontane-Schule, Leverkusen, Germany

Higher Education

10/2002–04/2009	Master Degree (Dipl.-Ing.) in Electrical Engineering, RWTH Aachen University
09/2005–02/2006	Erasmus-Semester at Universidad Politécnica de Catalunya (UPC), Barcelona, Spain

Professional Experience

04/2009–05/2015	Research Assistant at the Institute of Technical Acoustics, RWTH Aachen University
04/2008–09/2008	Intern at the Institute for Research in Construction (IRC), National Research Council (NRC), Ottawa, Canada
10/2006–04/2008	Student assistant at the Institute of Technical Acoustics, RWTH Aachen University
07/2004–07/2005	Student worker in the IT department of the Philips Lighting Factory, Aachen

Bisher erschienene Bände der Reihe

Aachener Beiträge zur Technischen Akustik

ISSN 1866-3052

Alle erschienenen Bücher können unter der angegebenen ISBN-Nummer direkt online
(http://www.logos-verlag.de) oder per Fax (030 - 42 85 10 92) beim Logos Verlag
Berlin bestellt werden.